T0314165

Cloud Ethics

Cloud

Louise Amoore

Ethics

Algorithms
and the Attributes
of Ourselves
and Others

Duke University Press
Durham and London
2020

© 2020 Duke University Press
All rights reserved
Printed and bound by CPI Group (UK) Ltd, Croydon, CR0 4YY
Designed by Drew Sisk
Typeset in Portrait Text, SangBleu Kingdom, and Helvetica Neue
by Copperline Book Services

Library of Congress Cataloging-in-Publication Data
Names: Amoore, Louise, author [date].
Title: Cloud ethics : algorithms and the attributes of ourselves
and others / Louise Amoore.
Description: Durham : Duke University Press, 2020. | Includes
bibliographical references and index.
Identifiers: LCCN 2019033257 (print) | LCCN 2019033258 (ebook) |
ISBN 9781478007784 (hardcover) | ISBN 9781478008316 (paperback) |
ISBN 9781478009276 (ebook)
Subjects: LCSH: Cloud computing—Moral and ethical aspects. |
Decision making—Moral and ethical aspects. | Algorithms—Moral
and ethical aspects.
Classification: LCC QA76.585 .A46 2020 (print) |
LCC QA76.585 (ebook) | DDC 004.67/82—dc23
LC record available at https://lccn.loc.gov/2019033257
LC ebook record available at https://lccn.loc.gov/2019033258

Cover art: <to come>

For my family, with love
&
For my students, with thanks for their curiosity

Any true scientific knowing is always, like feeling, only partial.
    —John Fowles

The possibility of the ethical lies in its impossibility.
    —Drucilla Cornell

CONTENTS

xi   *Acknowledgments*

1   *Introduction*
     Politics and Ethics in the Age of Algorithms

Part 1. Condensation

29   1   The Cloud Chambers:
       Condensed Data and Correlative Reason

56   2   The Learning Machines:
       Neural Networks and Regimes of Recognition

Part 2. Attribution

85   3   The Uncertain Author:
       Writing and Attribution

108   4   The Madness of Algorithms:
       Aberration and Unreasonable Acts

Part 3. Ethics

133   5   The Doubtful Algorithm:
       Ground Truth and Partial Accounts

154   6   The Unattributable:
       Strategies for a Cloud Ethics

173   *Notes*

197   *Bibliography*

212   *Index*

ACKNOWLEDGMENTS

The writing of this book coincided with what has felt like the most difficult and dangerous of political times I have experienced. The UK stands on the brink of leaving the community of European peoples that it joined in my first year of life, and far-right political forces are gathering at every turn. So many of the things in the world that I hold dear seem to be under threat: the right of all young people to have an education funded by those who have already benefited in the same way; the obligation on all people and governments to welcome and support those who flee from violence and hardship; the capacity of a democratic society to challenge hate and prejudice; the rights of women, black and ethnic minorities, and LGBTQ+ people to make political claims in the world without risk to their lives and well-being. Following the rising powers of algorithms, as they have begun to dominate economic and political life, my fears for the future are amplified. Let me be clear, machine learning algorithms that anticipate our future propensities are seriously threatening the chances that we have to make possible alternative political futures. In this context, the writing of this book has been challenging for its constant reminder of the limiting of potential political futures. But it has also been intensely uplifting for the incredible generosity of others in encouraging my work. One of the overriding feelings I have as I complete this book is that young people have the full measure of what is at stake ethicopolitically in an age of algorithms. After every lecture I have given, over every discussion in corridors and cafes, I have been struck by the depth and force of the understanding and innovative thought of the next generation of young scholars and students. There is much to be hopeful for, and I dedicate this book to this sense of hopefulness and curiosity.

I have worked in the Department of Geography at Durham University for fourteen years. It is not easy to describe how a group of people comes to matter so greatly to whether it will be possible to conduct the work one most wishes to. I can only say that the mood of experimentation, the quiet support of curious juxtapositions of theory and method, the standing together on a snowy union picket line, and the quite breathtaking generosity in reading work has been vital to me. Thank you all—you are pretty incredible: Ben Anderson, Andrew Baldwin, Oliver Belcher, Mike Bentley, Harriet Bulkeley, Sydney Calkin, Kate Coddington, Rachel Colls, Mike Crang, Niall Cunningham, Jonny Darling, Alex Densmore, Nicky Gregson, Paul Harrison, Sarah Hughes, Elizabeth

Johnson, Sarah Knuth, Paul Langley, Lauren Martin, Cheryl McEwan, Colin McFarlane, Joe Painter, and Jeremy Schmidt. The past few years have also been extraordinary for the PhD students I have been so lucky to supervise: Alice Cree, Arely Cruz Santiago, Andrei Dinca, Pete Forman, Sarah Hughes, Olivia Mason, Anna Okada, and Vanessa Schofield. I have never known any period so exciting for the discussions you have provoked and the experimental methods you are pioneering. You all have so much courage, and you make me feel more courageous.

When I reflect on the many forks in the path of writing this book, I can remember quite clearly how particular questions from an audience, or comments over lunch at a workshop, have opened avenues I had not imagined. I am immensely grateful to the many people who have so generously hosted me and engaged with my work. There are too many debts to others to recount fully here, but I thank you profoundly, and I will signal some of the branching points of my journey. A visit I made to the University of Wisconsin-Madison, at the invitation of Eric Schatzberg, Alan Rubel, and their colleagues, had a lasting effect on my work. The lecture audiences and workshop participants at the Holtz Center for Science and Technology Studies and the International Information Ethics Conference have influenced my thought more than they could possibly know—thank you. My time spent at Australian National University, Canberra, brought conversations that have endured—thank you, Fiona Jenkins and colleagues, for inviting me during your Divided Authorities theme, and Fiona and William for hosting my visit. For discussions on robotics, point clouds, and machine learning, I thank Jutta Weber, Oliver Leistert, and their colleagues at University of Paderborn, Germany. Thanks to Nanna Thylstrup and her collaborators on the Uncertain Archives project at University of Copenhagen, and to discussions with Luciana Parisi, Ramon Amaro, and Antoinette Rouvroy—for experiments with algorithms and archives. Peter Niessen and Christine Hentschel at the University of Hamburg probably did not realize how much their invitation to give a lecture in their Revenge of the Bots series would indelibly mark the final stages of my writing; thank you, generous audience, especially the computer scientists. Marc Boeckler, Till Straube, and their colleagues at Goethe University, Frankfurt, generously invited me as part of their Digital Disorders theme and made possible the most inspiring interdisciplinary conversations. Thank you. Sometimes, the process of writing can be a lonely affair, but these past years have felt full of conversation and inspiration. I am grateful for the many times I have felt that my work reached new audiences and different perspectives: thank you, Peter Burgess and his students at the École Nationale Supérieure, Paris; Pip Thornton,

Andrew Dwyer, Joe Shaw, and others present in the audience at the Turing Institute, London; Jeroen de Kloet and colleagues at the Amsterdam Centre for Globalisation Studies (and Erna Rijsdijk and Mike Shapiro for asking questions that I can still recall); Nathaniel Tkacz, Celia Lury, and their colleagues at the Warwick Centre for Interdisciplinary Methodologies for discussions of data streams; Huub Dijstelbloem and the University of Amsterdam's Politics of Visualization group; Martin Kornberger, Ann-Christina Lange, and their Copenhagen Business School colleagues working on digital infrastructures; Gunes Tavmen and her colleagues at the University of London, Birkbeck, and the Vasari Institute, for conversations on cloud ethics; Susan Schuppli, Luciana Parisi, Jussi Parrika, and Eyal Weizman, for memorable discussions of the architecture of algorithms; N. Katherine Hayles and all the participants in the Thinking with Algorithms conference in Durham; Jamie Woodward, Maja Zefuss, Noel Castree, Martin Coward, Veronique Pin-Fat, and those in the audience at the Manchester University 125 Years Anniversary lecture. You have no idea how much your generosity has supported my work.

I am grateful for the patience and time afforded to me by many people whose daily working lives bring them close to the lifeworlds of algorithms. Without the insights of computer scientists, doctors, software developers, lawyers—and many others who were willing to share their accounts—it would not have been possible to pursue my questions. It can be difficult to engage with new vocabularies and practices, but this kind of difficulty has been so productive in following threads in the dark and finding insight in the multiple meanings of bias, attribute, decision, weight, and judgment. It seems to me that the basis for ethics in an age of algorithms is not strictly a shared language and common understanding but precisely the persistence of difference and the inexplicable.

The writing of this book was funded by the award of a Leverhulme Major Research Fellowship, "Ethics of Algorithms" (2016–2018). I have been extraordinarily fortunate to be afforded the time and resources to do the work that I so wanted to be able to do. It has become much more difficult in the UK to fund the aspects of scholarship that involve reading, thinking, presenting and discussing work, and writing. Thank you, Leverhulme Trust, and may you long continue to act against the grain of prevailing logics of research funds in our times. During my fellowship, two extraordinary women have undertaken my teaching duties: Sarah Hughes and Sydney Calkin. They are a tour de force, and I thank them for everything they have done. Much of the fieldwork research for the book was funded by the Economic and Social Research Council (ESRC) for research on "Securing against Future Events" (2012–16). Throughout the project I worked in collaboration with Volha Piotukh—a postdoctoral researcher of simply astounding talent. Volha, my thanks will always fall short. You simply never gave up—you got us into places for interviews and observations that I would have said were impossible. You asked questions that challenged, exposed, and opened new pathways. You are the human embodiment of a whole new methodology for the social sciences and humanities to engage science. Hanging out with you (long-haul flights, broken trains) has been inspirational.

I owe an immeasurable debt to Duke University Press editor Courtney Berger, who has once more offered encouragement and thoughtful critique in equal measure. Thank you to Jenny Tan and Sandra Korn for guiding me patiently through the editorial and production process, and Lalitree Darnielle for working on the art and illustrations with such care. I am especially grateful to the three anonymous reviewers for the incredible investment of their time and insight in making this a better book. A version of chapter 1 appeared previously as "Cloud Geographies: Computing, Data, Sovereignty," *Progress in Human Geography* 42, no. 1 (2018): 4–24.

Finally, friends and family. You know who you are, and you know how you tolerated the tunnel vision that can be necessary to see something through. Marieke de Goede, we have walked and talked in countless places from Turin to San Diego over twenty-two years, and every conversation has left its mark. My gratitude to you is distilled to a dinner in Boston in April 2016 when you put the whole show back on the road—thank you. Jonathan, you listened patiently to all the nonsense and supplied a different kind of distillation, craft gin—a finer brother there cannot be. Grace and Tom—you are older now, and though the words "mum is finishing a book" bring necessary shrugs of the shoulders (and some giggling), you are so very kind about it. Paul, I only realized in these past few weeks that we never work on completing manuscripts at the same time. Like everything in our lives, they are shared across time, with love and support. Thank you for making it possible, and it is your turn again now.

Politics and Ethics
in the Age of Algorithms

The mathematical proposition has been given the stamp of incontestability.
I.e.: "Dispute about other things; this is immovable—it is a hinge on which your
dispute can turn."
  —Ludwig Wittgenstein, *On Certainty*

"A Hinge on Which Your Dispute Can Turn"

It is March 2016, and I am seated in a London auditorium, the gray curve of the
river Thames visible from the windows. A tech start-up business, specializing
in developing machine learning algorithms for anomaly detection, is presenting its latest algorithmic innovations to the assembled government and corporate clients. The projection screen displays a "protest monitoring dashboard"
as it outputs risk scores of "upcoming threats of civil unrest" in cities around
the globe, their names scrolling: Chicago, London, Paris, Cairo, Lahore, Islamabad, Karachi. The score that the analyst reads from the dashboard is the
singular output of deep neural network algorithms that have been trained to
recognize the attributes of urban public life, the norms and anomalies latent
in the data streams extracted from multiple sources, from Twitter and Facebook to government databases. As the presenter explains to the audience of
national security, policing, and border officials, "We train our algorithm to
understand what a protest is and is not," and "it gets better," "adapting day by
day," as it iteratively learns with humans and other algorithms.[1] The process of
learning "what a protest is" from the clustered attributes in data and modifying the model continues when the algorithm is later deployed in the city or at
the border: "We give you the code," he pledges, "so that you can edit it." How
does an algorithm learn to recognize what a protest is? What does it mean to
cluster data according to the attributes and propensities of humans to gather
in protest or in solidarity? At the London event, as the presenter displays a still

Figure I.1   An image from Stephen Spielberg's film adaptation of Philip K. Dick's novel *Minority Report* appears in a technology company's presentation to government and corporate analysts. Author's photograph.

from *Minority Report* (figure I.1), my thoughts turn to the protests that took place one year earlier, in the US city of Baltimore.

On April 12, 2015, Freddie Gray, a twenty-five-year-old African American man, sustained fatal injuries in the custody of the Baltimore Police Department. The profound violence of Gray's murder is an all-too-familiar event in the racialized architectures of our contemporary cities. During the days that followed his death, however, as people gathered on Baltimore's streets to protest the killing, the violence of the act extended into the plural actions of a set of machine learning algorithms that had been supplied to the Baltimore Police Department and the US Department of Homeland Security by the tech company Geofeedia. With the use of techniques similar to those described in the London protest-monitoring software, the Geofeedia algorithms had been trained on social media data, analyzing the inputs of Twitter, Facebook, YouTube, Flickr, and Instagram and producing scored output of the incipient propensities of the assembled people protesting Gray's murder. "Several known sovereign citizens have begun to post social media attempting to rally persons to demonstrate," recorded the Baltimore Police Department in a memo that promised to "continue to evaluate threat streams and follow all actionable leads."[2] Indeed, Geofeedia went on to market its algorithms to other states on the basis of a Baltimore "case study" (figure I.2) in which Freddie Gray is said to have "passed away," the city to have "braced itself for imminent protests," and the police to have seized "opportunities" to analyze "increased chatter from high school kids who planned to walk out of class."[3]

During those days in April, terabytes of images, video, audio, text, and biometric and geospatial data from the protests of the people of Baltimore were rendered as inputs to the deep learning algorithms. Even the written text embedded within social media images—such as the "police terror" placards carried aloft and captured on Instagram—was extracted by a neural network and became features in the algorithm.[4] People were arrested and detained based on the outputs of a series of algorithms that had—as the London scene also proposed—learned how to recognize what a protest is, what a gathering of

Baltimore County Police Department and Geofeedia Partner to Protect the Public During Freddie Gray Riots

BACKGROUND

When Freddie Gray passed away in Baltimore on April 25, 2015 from injuries allegedly sustained during his arrest by the City of Baltimore Police, Detective Sergeant Andrew Vaccaro with the Baltimore County Police Department's Criminal Intelligence Unit knew trouble was brewing. With Ferguson's Michael Brown still fresh in the nation's mind and racial tensions running high, Baltimore braced itself for the imminent and expected protests.

OPPORTUNITY

"The Freddie Gray incident was a watershed moment for the City of Baltimore police," Vaccaro said. "The minute his death was announced, we knew we needed to monitor social media data at key locations where protesting was likely, especially at the local police precinct where Gray had been arrested."

In a stroke of luck, the Baltimore County Police Department had renewed their Geofeedia contract a week before the trouble began. The Criminal Intelligence Unit had experienced the tool's power first-hand before, and they didn't hesitate to call in reinforcements when trouble arose.

When an event at Camden Yards on April 25 turned violent, a ten-night-long police nightmare was set into motion. It was the Crimi-

perimeters around key locations, set up automated alerts, and forward real-time information directly to Vaccaro's team via email

Figure I.2   Geofeedia's account of the Baltimore protests in the marketing of software analyzing social media data for the detection of incipient public protests. American Civil Liberties Union, 2016.

people in the city might mean. As Simone Browne has argued in her compelling account of the "digital epidermalization" of biometric algorithms, what is at stake is the recognizability of a body as human, as fully political.[5] Among Baltimore's arrests and detentions were forty-nine children, with groups of high school students prevented from boarding buses downtown because the output of the algorithm had adjudicated on the high risk they posed in the crowd.[6] Based on the so-called ground truth of features that the algorithms had learned in the training data, the algorithms clustered the new input data of people and objects in the city, grouping them according to their attributes and generating a numeric scored output.[7]

The profound violence of the killing of one man, and the residue of all the past moments of claims made in his name, and in the name of others before him (note that the names Freddie Gray and Michael Brown persist in the training of subsequent algorithms to arbitrate protest), becomes lodged within the algorithms that will continue to identify other faces, texts, and signs in future crowds. Understood as the principal architecture of what N. Katherine Hayles calls the "computational regime," what matters to the algorithm, and what the algorithm makes matter, is the capacity to generate an actionable output from a set of attributes.[8] What kind of new political claim, not yet registered as claimable, could ever be made if its attributes are recognizable in advance? The very capacity to make a political claim on the future—even to board a bus to make that claim—is effaced by algorithms that condense multiple potential futures to a single output.

At the level of the algorithm, it scarcely matters whether the clustered attributes are used to define the propensities of consumers, voters, DNA sequences, financial borrowers, or people gathering in public space to make a political claim.[9] Thus, when in 2016 Cambridge Analytica deployed its deep learning algorithms to cluster the attributes of voters in the UK EU referendum and the US presidential election, or when Palantir's neural networks supply the targets for the US ICE deportation regime, what is at stake ethicopolitically is not only the predictive power of algorithms to undermine the democratic process, determine the outcomes of elections, decide police deployments, or make financial, employment, or immigration decisions. Of greater significance than these manifest harms, and at the heart of the concerns of this book, algorithms are generating the bounded conditions of what a democracy, a border crossing, a social movement, an election, or a public protest could be in the world.

Ethics of Algorithms

At first sight, the potential for violent harm precipitated by algorithms that learn to recognize human propensities appears to be a self-evident matter for critique. Surely, one could say, the ethical terrain of the algorithm resides in the broader political landscape of rights and wrongs, good and evil. After all, one could readily identify a set of rights, already apparently registered as belonging to rights-bearing subjects, that has been contravened by algorithms that generate targets, adjudicating which people may peaceably assemble, or which people are worthy of credit or employment, and on what terms. Indeed, on this terrain of delineating the rights and wrongs of algorithmic actions is precisely where many critical voices on the harms of the algorithm have been heard. Writing in the *New York Times*, for example, Kate Crawford identifies machine learning's "white guy problem," arguing that "we need to be vigilant about how we design and train machine learning systems."[10] The dominant critical perspectives on algorithmic decisions have thus argued for removing the "bias" or the "value judgements" of the algorithm, and for regulating harmful and damaging mathematical models.[11] Within each of these critical calls, the ethical problem is thought to dwell in the opacity of the algorithm and in its inscrutability, so that what Frank Pasquale has called the "black box society" is addressed through remedies of transparency and accountability.[12] In sum, the rise of algorithmic power in society has been overwhelmingly understood as a problem of opaque and illegible algorithms infringing or undercutting a precisely legible world of rights belonging to human subjects. In such a framing, there is an outside to the algorithm—an accountable human subject who is the locus of responsibility, the source of a code of conduct with which algorithms must comply. To call for the opening of the black box, for transparency and accountability, then, is to seek to institute arrangements that are good, ethical, and normal, and to prevent the transgression of societal norms by the algorithm.

Yet, when people gathered to protest on Baltimore streets, or when Facebook users' data fueled the political and commercial models of Cambridge Analytica (figure I.3), legible rights to peaceable assembly or to electoral due process were not violated primarily by illegible algorithms. Rather, the means by which people could appear in a political forum, the conditions of their appearance, and the capacities they had to make a recognizable political claim in the world were subject to algorithmic regimes of what Michel Foucault calls truth telling and wrongdoing.[13] In short, what matters is not primarily the identification and regulation of algorithmic wrongs, but more significantly how algo-

Figure I.3   Cambridge Analytica advertises how "data drives all we do" in the fused commercial and political pursuit of ways "to change audience behavior." Screenshot archived by the author in May 2018, when the firm ceased trading under that name.

rithms are implicated in new regimes of verification, new forms of identifying a wrong or of truth telling in the world. Understood in these terms, the algorithm already presents itself as an ethicopolitical arrangement of values, assumptions, and propositions about the world. One does not need to look beyond the algorithm for an outside that is properly political and recognizably of ethics. Indeed, there can be no legible human outside the algorithm and underwriting its conduct, for as John Cheney-Lippold reminds us, we are enmeshed in the data that produce each "freshly minted algorithmic truth."[14] One cannot sustain a search for codes of ethics that instill the good, the lawful, or the normal into the algorithm. Contemporary algorithms are not so much transgressing settled societal norms as establishing new patterns of good and bad, new thresholds of normality and abnormality, against which actions are calibrated.

Actions one might consider harmful, as William Connolly notes, are not merely "actions by immoral agents who freely transgress the moral law" but are "arbitrary cruelty installed in regular institutional arrangements taken to embody the Law, the Good, and the Normal."[15] Amid the widespread search for new ethical arrangements for the scrutiny and regulation of algorithms, what becomes of the arbitrary harms lodged within embodied algorithmic arrangements? One could imagine a world in which the deep neural networks used in cities like Baltimore are scrutinized and rendered compliant with rules and yet continue to learn to recognize and misrecognize people and to infer intent, to generate rules from the contingent and arbitrary data of many past moments of associative life on the city streets, to refine and edit the code for future uses in unknown future places. I may feel that some notion of legible rights is protected, and yet the attributes generated from my data, in correlation with yours and others', continue to supply the conditions for future arbitrary actions against unknown others. I draw a distinction here between ethics as code, or what Michel Foucault describes as "the code that determines which acts are permitted or forbidden," and ethics as the inescapably political formation of the relation of oneself to oneself and to others.[16] My argument is that there is a need for a certain kind of ethical practice in relation to algorithms, one that does not merely locate the permissions and prohibitions of their use. This different kind of ethical practice begins from the algorithm as always already an ethicopolitical entity by virtue of being immanently formed through the relational attributes of selves and others. My desire for a different mode of critique and ethical account is animated not by the question, How ought the algorithm be arranged for a good society?, but by the question, How are algorithmic arrangements generating ideas of goodness, transgression, and what society ought to be?

In this book I propose a different way of thinking about the ethicopolitics of algorithms. What I call a *cloud ethics* is concerned with the political formation of relations to oneself and to others that is taking place, increasingly, in and through algorithms. My use of the term *cloud* here is not confined to the redefined sovereignties and technologies of a "cloud computing era," as understood by Benjamin Bratton and others, but refers to the apparatus through which cloud data and algorithms gather in new and emergent forms.[17] The *cloud* in my cloud ethics is thus closer to that envisaged by John Durham Peters, for whom clouds are media in the sense that they are "containers of possibility that anchor our existence and make what we are doing possible."[18] To consider algorithms as having ethics in formation is to work with the propensities and possibilities that algorithms embody, pushing the potentials of their arrangements beyond the decisive moment of the output.

A cloud ethics acknowledges that algorithms contain, within their spatial arrangements, multiple potentials for cruelties, surprises, violences, joys, distillations of racism and prejudice, injustices, probabilities, discrimination, and chance. Indeed, many of the features that some would like to excise from the algorithm—bias, assumptions, weights—are routes into opening up their politics. Algorithms come to act in the world precisely in and through the relations of selves to selves, and selves to others, as these relations are manifest in the clusters and attributes of data. To learn from relations of selves and others, the algorithm must already be replete with values, thresholds, assumptions, probability weightings, and bias. In a real sense, an algorithm must necessarily discriminate to have any traction in the world. The very essence of algorithms is that they afford greater degrees of recognition and value to some features of a scene than they do to others. In so doing, algorithms generate themselves as ethicopolitical beings in the world. If to have ethics is not merely to have a code prohibiting, for example, bias or assumptions, but to work on oneself via relations, then the ethicopolitics of algorithms involves investigations of how they learn to recognize and to act, how they extract assumptions from data relations, and how they learn what ought to be from relations with other humans and algorithms.

To be clear, the cloud ethics I propose here does not belong to an episteme of accountability, transparency, and legibility, but on the contrary begins with the opacity, partiality, and illegibility of all forms of giving an account, human and algorithmic. To advance a cloud ethics is to engage the ungrounded politics of all forms of ethical relations. The significant new ethical challenges that algorithms seem to present to society actually manifest novel features of some profoundly old problems of the grounds for ethical action. As Judith Butler explains in her Spinoza lectures, the demand to give an account of oneself will always fall short, for "I cannot give an account of myself without accounting for the conditions under which I emerge."[19] If one assumes that the determination of an unequivocal *I* who acts is a necessary precondition of ethics, as Butler cautions, then this identifiable self is "dispossessed" by the condition of its emergence in relation to others. For Butler, this persistent failure to give a clear-sighted account does not mark the limit point of ethics. On the contrary, the opaque and unknowable nature of making all kinds of acting subjects is the condition of possibility of having an ethicopolitical life.[20]

In short, and in contrast to the equation of ethics with transparency and disclosure, ethical responsibility is sustained by conditions of partiality and opacity. My notion of a cloud ethics extends the opacity of the human subject, envisaging a plurality of venues for ethical responsibility in which all selves—human and algorithmic—proceed from their illegibility. The apparent opacity and illegibility of the algorithm should not pose an entirely new problem for human ethics, for the difficulty of locating clear-sighted action was already present. The *I* who forms the ethical relation was always in question and is now, with algorithms, in question in new ways. Though the mathematical propositions of algorithms cannot be made fully legible, or rendered accountable, they can be called to give accounts of the conditions of their emergence. These conditions include some relations that are identifiably between humans and algorithms—such as the selection and labeling of training data, the setting of target outputs, or the editing of code "in the wild," for example—but others still are relations of algorithms to another algorithm, such as a classifier supplying the training data from which a neural network will learn. In all such instances of iterative learning, the significant point is that the conditions of an algorithm's emergence—a composite of human-algorithm relations—are venues for ethicopolitics.

In a discussion on the impossibility of the transparent algorithm, the brilliant and generous scholar of black studies and machine learning Ramon Amaro once said, "Well what *would* it be if we even *could* open it? It's just math."[21] Of course, he intended the comment as a provocation, but mathematics is never only "just math," as Amaro's work vividly shows. To reflect on the conditions of an algorithm's emergence is also to consider how, as mathematical knowledge forms, algorithms have achieved the status of objective certainty and definiteness in an uncertain world.[22] Ludwig Wittgenstein observed mathematical propositions to be "given the stamp of incontestability," a mark of the "incontrovertible" and an "exemption from doubt" that other propositions, such as "I am called," are not afforded.[23] For Wittgenstein, mathematics as practice—like all other language games—is concerned with particular uses of propositions, where "what a proposition is, is in one sense determined by the rules of sentence formation, and in another sense by the use of the sign in the language game."[24] His concern is that the mathematical proposition has achieved a particular status of certainty in an otherwise uncertain world, so that it becomes "a hinge on which your dispute can turn."[25] For Wittgenstein, the mathematical proposition should be regarded as no less doubtful or uncertain than the "empirical propositions" otherwise made about the world. Indeed, Wittgenstein's point is to address mathematical propositions *as empirical actions* that are "in no way different from the actions of the rest of our lives, and are in the same degree liable to forgetfulness, oversight and illusion."[26] Following Wittgenstein, the use of mathematical propositions is profoundly social and, in my reading, ethicopolitical. An algorithm is formulated through a series of truth claims about its match to the world, and yet, in its use in the world it is as prone to forgetfulness, oversight, misrecognition, and illusion as any other language game.

Algorithms such as those used to detect latent social unrest in the city may appear in the world as Wittgenstein's "hinge on which your protest can turn" in the most direct sense that the hinge delimits and circumscribes the arc of the politics of protest. But the algorithm as hinge does not merely mark the limit point of resistance; rather, it presents something as a singular optimal output, when it is actually generated through multiple and contingent relations. My cloud ethics considers the algorithmic hinge to be akin to Karen Barad's scientific apparatus, which decides what matters in the world, what or who can be recognized, what can be protested, and which claims can be brought.[27] Understood in this way, the algorithm is not the hinge as an incontrovertible axis, exempted from doubt, on which all social, political, and economic life turns. "The hinge point," as Foucault differently identifies, can also be the point of "ethical concerns and political struggle," as well as the point of "critical thought against abusive techniques of government."[28]

So, a principal ethicopolitical problem lies in the algorithm's promise to render all agonistic political difficulty as tractable and resolvable. Where politics expresses the fallibility of the world and the irresolvability of all claims, the algorithm expresses optimized outcomes and the resolvability of the claim in the reduction to a single output. In the following pages, I specify how this book approaches what an algorithm is and what it does. Among the many problems of studying algorithms is the matter of specifying which type of algorithm one is addressing. Though my primary focus in this book is on machine learning algorithms, and predominantly deep neural networks, in most of the instances I discuss, these algorithms are used in conjunction with some much older and less fashionable rules-based and decision tree algorithms. The form of the algorithm is not delimited by its name but by its coming into being, its use in the wild. As Nick Seaver has argued, "Rather than offering a 'correct' definition," a critical study of algorithms could begin from "their empirical profusion and practical existence in the wild."[29] For example, an advanced deep neural network for object recognition is intimately connected to some much older classifiers that are used in the preparation of the data, and it meaningfully comes into being as it is modified through its deployment in the world. It is not possible to identify a secure boundary around one specific named algorithm because, as a calculative device, it is a composite creature. I propose three routes into understanding what algorithms are in the world, each with its own distinctive implications for what is at stake ethicopolitically: algorithms as *arrangements of propositions*; algorithms as *aperture instruments*; algorithms as *giving accounts of themselves*.

"Generate All the Rules and Debate Which Is Interesting": Algorithms as Arrangements of Propositions

The spatiality of the logic of algorithms is most commonly figured as a series of programmable steps, a sequence, or a "recipe," governed by "precise instructions" within a "finite procedure."[30] The calculative logic is thus most widely represented as a logic of the series, where each step in a calculative procedure is defined by its position in a finite series.[31] This notion of a recipe, or a series of steps, contributes to limiting the imagination of what the ethicopolitics of algorithms could be. The spatial imaginary of the algorithm as a series of steps nurtures a particular set of ideas about how to intervene in the series to change the outcomes.[32] So, for example, if the negative outcome of a credit-checking algorithm was found to include a racial or gendered bias in one of the steps, then the removal of this element in the recipe could be considered significant in notions of accountable and responsible algorithmic decisions. Similarly, to envisage a "kill switch" in the algorithms of autonomous weapons systems, for example, is to imagine a sequence of steps in which a human in the loop could meaningfully intervene and prevent a lethal decision.[33] Where algorithms are represented as a sequence, or what Manuel DeLanda calls "mechanical recipes specified step by step," the addition or deletion of a step destroys the outcome, halts the decision in its tracks.[34] In this sense the spatial imagination of the algorithm as series appears to make possible all kinds of human oversight of otherwise automated decisions.

The representation of algorithms as a logical series, however, seriously overlooks the extent to which algorithms modify themselves in and through their nonlinear iterative relations to input data. The machine learning algorithms that are so categorically redefining our lives are characterized less by the series of steps in a calculation than by the relations among functions. Within computer science these relations are understood to be recursive functions, whereby the output of one calculation becomes the defining input for another, and so on, with each function nested within others like an infinite nesting of Russian dolls. These recursive functions, as Paulo Totaro and Domenico Ninno have argued, are having specific and durable effects on contemporary society.[35] Significantly, the removal or deletion of one function does not destroy the overall arrangement. Indeed, intrinsic to the logic of machine learning algorithms is their capacity to learn which outputs from which of their layers to pay greater attention to and which to bypass or discard.[36] This matters greatly, because the removal of a step one assumed to contain sensitive data on race, for example, would not remove or delete the process of learning via proxies how to recognize by means of racialized attributes. Consider, for example, the machine learning algorithms that are being used in sentencing decisions by the courts to anticipate the optimal outcome of a prison sentence versus noncustodial measures.[37] Such technologies do not deploy a sequential logic that could be amenable to oversight and intervention in the steps. On the contrary, the optimal future outcome is defined solely in relation to an array of recursive functions, a "different mode of knowing," as Adrian Mackenzie explains, in which the input data on past convictions and sentencing outcomes of hundreds of thousands of unknown others supply the contingent probabilities to all the layers within the algorithm.[38]

Algorithms are not merely finite series of procedures of computation but are also generative agents conditioned by their exposure to the features of data inputs. As Luciana Parisi has argued, this is "a new kind of model," which "derives its rules from contingencies and open-ended solutions."[39] When algorithms learn by inductively generating outputs that are contingent on their input data, they are engaging experimentally with the world. As computer scientist Rakesh Agrawal explains the shift from rules-based to generative learning algorithms, past forms "used a statistical notion of what was interesting" so that "the prevailing mode of decision making was that somebody would make a hypothesis, test if it was correct, and repeat the process." With machine learning algorithms, such as recursive neural networks, Agrawal notes that "the decision process changed," and algorithms would "generate all rules, and then debate which of them was interesting."[40] In their contemporary form, algorithms generate output signals that open onto uncertainty as to what is interesting, useful, or optimal. These output signals are not mere mathematical abstractions but are actionable propositions, such as "this person poses high risk of overstaying their visa," or "this object is threatening the security of the street." Often, when algorithm designers describe how they work with clients on a particular application, they suggest that they "tune" the algorithm as part of a discussion with the client of what is useful or optimal. This experimental tuning enacts the process Agrawal describes as debating which of the outputs is interesting, where the observation of the output of the model modifies and adjusts the weightings and thresholds of the algorithm. A kind of science of emergent properties, as I describe in my book *The Politics of Possibility*, such techniques significantly transcend and undercut traditional statistical notions of what matters, what is interesting, and what is optimal.[41]

In this book I understand the spatial logic of algorithms to be an *arrangement of propositions* that significantly generates what matters in the world. In contrast to the spatiality of the series or recipe, the arrangement of propositions articulates the algorithm's capacity to engage experimentally with the world, to dwell comfortably with contingent events and uncertainties, and yet always to be able to propose, or output, an optimal action. Practically, in the research for this book, I have studied the algorithm not as a finite series of programmable steps but as perennially adjustable and modifiable in relation to a target output. What does it mean to understand algorithms as arrangements of propositions? In Alan Turing's famous paper "Systems of Logic Based on Ordinals," he reflected on what he called "the exercise of ingenuity in mathematics." For Turing, ingenuity was important for mathematical reasoning because it provided "arrangements of propositions," which meant that the intuitive mathematical steps could not "seriously be doubted."[42] The arrangement of propositions was thus a kind of mathematical architecture that supported the intuitive and the inferential faculties. As I use the notion, an arrangement of propositions extends to the experimental and iterative capacities of algorithms to propose things in and about the world. This is most likely not an interpretation Turing would approve of. Indeed, when he attended Wittgenstein's 1939 Cambridge lectures, their manifest disagreements in the lecture theater concerned precisely the matter of whether mathematical propositions could have normative effects. When Turing asserted that "from the mathematical theory one can make predictions," Wittgenstein replied that "pure mathematics makes no predictions." For Wittgenstein, "30 × 30 = 900 is not a proposition about 30" but rather a proposition that finds its expression only in the "grammar" of its arrangement.[43] While for Turing, the numeric value 30 itself has a kind of agency, for Wittgenstein this is afforded only by its arrangement in a wider grammar through which it comes into use.

In a sense, the disagreement between Turing and Wittgenstein is undercut by a twenty-first-century world in which algorithms arguably generate their grammars and propositions through their exposure to numbers in the form of input data. As historian of mathematics Keith Devlin reminds us, mathematics involves not only numeric values based on a "count" but also, crucially, transformations based on "processes you perform."[44] The arrangements of propositions I envisage are not only numeric but transformative and performative. They contain within them multiple combinatorial possibilities and connections. The multiplicity of the algorithm—its plural possible combinations, pathways, weights, and connections—matters greatly to my tracing of the empirical processes by which algorithms learn and reach decisions, and to my desire to shift the ethicopolitical terrain on which this is understood to take place. The arrangement of propositions means that an apparently optimal output emerges from the differential weighting of alternative pathways through the layers of an algorithm. In this way, the output of the algorithm is never simply either true or false but is more precisely an effect of the partial relations among entities. As Isabelle Stengers has noted of the proposition, "It is crucial to emphasize that the proposition in itself cannot be said to be true or false" because in itself "it is indeterminate with regard to the way it will be entertained."[45] As a proposition, the algorithm can similarly not be said to be true or false—it cannot be held to account for its relation to truth in this sense. A pattern of false positives from a biometric algorithm, for example, can never be simply false because the threshold is immanently adjustable. Understood in Stengers's sense of the proposition, algorithms are "indeterminate with regard to the way [they] will be entertained."

Consider, for example, the multiple arrangements of propositions as they are manifest in the Asimov Institute's Neural Network Zoo. Described as an "almost complete chart of neural networks," the neural net zoo displays a spatial mapping of the arrangements of propositions of machine learning algorithms. This is not a zoo that categorizes its flora and fauna by the characteristics of their genus and species. Instead, it displays the algorithms' architectures as arrangements of proximities, distances, intensities, and associations. The Asimov researchers, in their depiction of the arrangement of a convolutional neural network (CNN)—commonly used for facial recognition, feature extraction, and image classification—explain how each node concerns itself only with its close neighboring cells. The nature of the function performed within the node is decided by the close communion of weighted probabilities in the neighboring cells. As an arrangement of propositions, one could not meaningfully open or scrutinize the 60 million probability weightings that make it possible for a CNN algorithm to recognize the attributes of a face in a crowd, declaring them to be true or false. Indeed, following Stengers's formulation of the proposition, the output of a facial recognition algorithm is never either "true" or "false" but instead is a useful proposition that can be infinitely recombined. Unlike a series of steps or a recipe, one could never have oversight of the infinite combinatorial possibilities of the algorithm as proposition. Nor would the deletion of a step render the whole unworkable. Once the algorithm is understood as an arrangement of propositions, the mode of ethics must work with the partiality and illegibility of the relations among entities.

"Reduced to That Which Interests You": Algorithms as Aperture Instruments

Critical accounts of the rise of algorithms have placed great emphasis on the power of algorithms to visualize, to reprogram vision, or indeed even to "see" that which is not otherwise available to human regimes of visuality. Similar to the spatial arrangement, this primacy of the visual register has also annexed what could count as the ethics and politics of algorithms. There are two curiously twinned accounts of contemporary algorithms in relation to regimes of sight and vision. The first is that algorithms operate on a plane in excess of human visibility and at scales that are inscrutable to the human. The second is that algorithms themselves have an enhanced capacity to visualize the invisible, to see, scan, and search volumes and varieties of data heretofore unavailable to human senses. Indeed, this intersection of machinic and human vision comes to the fore in the espoused ethics of public inquiries into the state's deployment of automated algorithms for the government of the population. For example, in the UK parliamentary inquiry following the Edward Snowden disclosures of widespread automated data analysis, a peculiar kind of virtue was found in the notion that, in machine learning intelligence, "only a tiny fraction of those collected are ever seen by human eyes."[46] Similarly in the United States, the former director of national intelligence James Clapper likened the NSA's algorithmic analysis of citizens' data to a form of library in which few books are "actually read" and where the output of the system supplies "the books that we need to open up and actually read."[47] There is an acute problem, then, with the widespread appeal to ethical codes that regulate what algorithms or humans do or do not see. It is a problem, I suggest, with its roots in the privileging of sight and vision over other forms of making things perceptible. "Vision cannot be taken," writes Orit Halpern in her wonderful book *Beautiful Data*, "as an isolated form of perception" but must be considered "inseparable from other senses."[48] To act and to be responsible for action, an algorithm need not "see" or "read" but need only make something or someone perceptible and available to the senses.

In this book I situate the ethics and politics of algorithms within a genealogy of technologies of perception. Contemporary algorithms are changing the processes by which people and things are rendered perceptible and brought to attention. This is definitively not merely a matter of making things amenable to vision and indeed is frequently a matter of sustaining something beneath the visual register and yet perceptible. As art historian Jonathan Crary writes, "Ideas about perception and attention were transformed" alongside the historical "emergence of new technological forms of spectacle, display, projection, attention, and recording."[49] Understood in this way, the transformation of perception involves changes in how the perceiving subject thinks about what could be brought to attention, changes in the horizon of possibility of human action. As with the advent of the technologies of printing press, camera, or cinema, so the advent of the machine learning algorithm implies a reworking of what it means to perceive and mediate things in the world.[50] This is not a process that is effectively captured by the idea that automated systems are undermining or superseding human forms of perception and action. To foreground instruments of perception, or what Henri Bergson terms "organs of perception," is to breach conventional distinctions between humans and machines and acknowledge the entangled nature of all forms of perception.[51] Bergson insists on the shared limits of perception across science and ordinary everyday experience, so that "ordinary knowledge is forced, like scientific knowledge," to divide up time into perceptible slices, to "take things in a time broken up into particles."[52] Whether the organ of perception is microscope, telescope, eye, camera, or algorithm, perception is attuned to action, to the dividing up of movement into points on a trajectory so that they can be acted on. "What you have to explain," he writes, is not "how perception arises, but how it is limited, since it should be the image of the whole, and is in fact reduced to the image of that which interests you."[53] Following Bergson's insight on how an organ of perception seizes the object of interest from its environment, to consider algorithms as instruments of perception is to appreciate the processes of feature extraction, reduction, and condensation through which algorithms generate what is of interest in the data environment.

Confronted by something of a moral panic surrounding the expansive volumes of "big data" and powers of surveillance of automated systems, my emphasis on practices of perception foregrounds precisely the opposing process: the reducing, distilling, and condensing of particles of interest from a whole. A defining ethical problem of the algorithm concerns not primarily the power to see, to collect, or to survey a vast data landscape, but the power to perceive and distill something for action. Algorithms function with something like an aperture—an opening that is simultaneously a narrowing, a closure, and an opening onto a scene. Let us consider, for example, an algorithm designer working in the UK defense sector, demonstrating the capacity of his deep neural network algorithms to recognize a mistaken civilian target amid a crowded data environment of drone images.[54] He shows a slice through time as a vehicle travels away from the center of Kandahar, Afghanistan. He explains the problem for the decision: that this could be a suspect vehicle or, crucially, a school bus taking children home to villages outside the city. The algorithms had learned to recognize a school bus through training data that supplied images of predominantly yellow US-style buses. The designer explains that his algorithm is aggregated with many others to generate a single actionable output—target/no target—but to do this he must necessarily reduce and condense the patterns of interest from a volume of input data. The training data—and the humans who labeled it—have conditioned the CNN algorithm to carve out and value some objects and to discard others. In this part of Afghanistan, some of the school buses are indeed in the spatial form of a US-style school bus, but others still are open-back trucks repurposed for transporting scholars. My point is that a potential act of violence, such as a school bus wrongfully targeted by a drone strike, resides not primarily in the vertical surveilling and collecting of data, but in fact in the horizontal arraying of possible patterns of interest lodged within the algorithm itself.[55] As an aperture instrument, the algorithm's orientation to action has discarded much of the material to which it has been exposed. At the point of the aperture, the vast multiplicity of video data is narrowed to produce a single output on the object. Within this data material resides the capacity for the algorithm to recognize, or to fail to recognize, something or someone as a target of interest.

The ethical stakes of what Mark Hansen calls "potential perceptual reconfiguration" applied to my cloud ethics necessarily involves something like a reopening of the process of the algorithm's reduction of a multiplicity to one.[56] What is happening in this process of condensing plural possible pathways to a single output? When an algorithm determines whether a vehicle is a military or a civilian target, or when it decides if a public protest contains latent dangerous propensities, it reduces the heterogeneity of durational time to perceive the attributes of an object and their differences of degree from other objects encountered in a past set of data. The question of what this crowd could be, what this vehicle might do, the frustrations or discomforts of the actual lived experience of waiting or gathering persist as indeterminacies in the hidden layers of the algorithm. Even within the archive of training data—sometimes just a Google ImageNet dataset of labeled images of school buses—are the residual contingencies of durational time, with all the past lived moments supplying norms and anomalies for the algorithm to learn. "My own duration, such as I live it in the impatience of waiting," reflects Gilles Deleuze, "serves to reveal other durations that beat to other rhythms, that differ in kind from mine."[57] To respond to the perceptual power of the algorithm and to prize open the aperture of the single output is to trace the other durations that continue to beat in the discarded data, the multiple other potential pathways that could be mapped between fragments.

"At the Limits of What One Knows": Algorithms Give Accounts of Themselves

Perhaps the most widespread concern in public and scholarly discourse on the operation of algorithms in society is that they are unaccountable or that they cannot meaningfully be held to account for harmful actions. Indeed, a kind of proxy form of accountability is emerging, in which the designers of algorithms are made the locus of responsibility for the onward life of their algorithms. Virginia Eubanks proposes a "Hippocratic oath for data science," in which the designers of algorithms would be accountable to human beings and not "data points, probabilities, or patterns."[58] Similarly, in a 2016 *Nature* editorial calling for "more accountability for big data algorithms," the editors propose that "greater transparency" could be achieved if the designers of algorithms "made public the source of the data sets they use to train and feed them."[59] They urge greater disclosure of the design of algorithms and an "opening up to scrutiny" of their workings. Indeed, so widespread is this notion that the accountability of algorithms can be grounded in their design or source code that technical solutions for "explainability" are being developed to trace an apparent bias back to a design problem. For example, automated systems for the assessment of creditworthiness are thought to be rendered transparent by a tool that traces the specific credit score output back to a data element, such as an unpaid bill.[60] Such techniques are thought to anchor the accountability of the algorithm precisely in an intelligible knowledge of its workings.

Within these demands for algorithmic accountability lies a specific form of giving an account. The locus of a truthful account is in the apparent "source" of the algorithm, in its origins, whether in the source code or in the algorithm designer as an author. This locus of original account, I suggest, is profoundly limiting the capacity to demand that algorithms give accounts of themselves. It imagines a secret source or origin to which all potential future harms could be traced. As Foucault writes on the disappearance of the author, "the task of criticism is not to bring out the work's relationship with the author" but rather to "analyse the work through its structure, its architecture, its intrinsic form, and the play of its internal relationships."[61] For our contemporary times, the call for accountability of algorithms has precisely targeted the work's relationship with the author, seeking to render transparent the intent and the workings of the design. At one level there are clear limits to identifying the source of the algorithm, not least that each apparent "one" contains multiple elements from multiple sources, with much of this aggregation concealed even from the designer. My point, however, is that the problem of anchoring accountability in a source or origin is not confined to algorithms but is a persistent and irresolvable ethicopolitical problem.

The problem of the unidentifiable origin of the algorithm extends to all notions of an authoring subject *I* who can give a clear-sighted account of herself. If one's account of oneself can never be fully secured, then the full disclosure of one's grounds for action is impossible. To give an account is always to give an uncertain narrative that risks falling short or failing to be recognized. Let us begin not from a search for secure grounds for accountability, then, but from the very ungroundedness of all forms of giving an account. "Ethics and politics only come into being," Thomas Keenan writes, "because we have no grounds, no reliable standpoints" from which to forge foundations.[62] To be responsible for something—an errant output, a fatal decision, a wrong judgment—is less a matter of securing the grounds for the action than a matter of responding even when knowledge is uncertain and the path is unclear. "What could responsibility mean," asks Keenan, if it is "nothing but the application of a rule or decision."[63] There is no great origin or source of responsibility without uncertainty and undecidability.

In this book I propose that algorithms are not unaccountable as such. At least, they should not be understood within a frame of ethical codes of accountability in which the source of the problem could be secured. Algorithms, I propose instead, are giving accounts of themselves all the time. These accounts are partial, contingent, oblique, incomplete, and ungrounded, but, as N. Katherine Hayles vividly documents, this is not a condition unique to the cognitive complexities of algorithms.[64] Far from it. The condition of giving an account that is never transparent or clear sighted is already the ethicopolitical condition with which we must live. To attend to the accounts that algorithms give is to "stay with the trouble" of "unexpected collaborations and combinations," as Donna Haraway has captured the "method of tracing, of following a thread in the dark."[65] Given the conditions of following collaborative threads in the dark, perhaps one should not create a special category of opaque and illegible agency for the identification of algorithms. If, as Judith Butler suggests, "my account of myself is partial, haunted by that for which I have no definitive story," then might it be that "the question of ethics emerges at the limits of our schemes of intelligibility, where one is at the limits of what one knows and still under the demand to offer and receive recognition?"[66] This is the ethics at the limits of schemes of intelligibility, of following threads in the dark, that I envisage for my cloud ethics. Refuting the many demands for an impartiality of the algorithm, excised of bias and prejudice, I wish to be alert to the always already partial accounts being given by algorithms.

As feminist scholars of technoscience have long reminded us, the partial account is not an account devoid of insight or real purchase on the world. "There is no unmediated photograph or passive camera obscura in scientific accounts of bodies and machines," writes Donna Haraway. "There are only highly specific possibilities, each with a wonderfully detailed, active, partial way of organizing worlds."[67] To attend to algorithms as generating active, partial ways of organizing worlds is to substantially challenge notions of their neutral, impartial objectivity. To foreground partiality is also to acknowledge the novel forms of distributed authorship that newly entangles the *I* who speaks in composite collaborations of human and algorithm. If one element of my past presence on a London street for a "Stop the War" campaign march enters a training dataset for a multinational software company's neural net, which, one day in the future, intervenes to detain some other person in a distant city, how is some part of my action lodged within this vast and distributed authorship? What is the possibility of my ethicopolitical responsibility for the dark thread to a future intervention made partially on the attributes of my past data?

To be attentive to the accounts algorithms are giving of themselves, then, is to begin with the intractably partial and ungrounded accounts of humans *and* algorithms. As is so manifestly present with the surgical robotics algorithms I discuss in chapter 2, the embodied accounts human surgeons give of what they can do, how they decide on boundaries between diseased and healthy tissue, how they can reach decisions, are not meaningfully separable from the surgical robots with whom they share a cognitive workload. The robot can only act to suture a wound because its algorithms have learned from the past data of many thousands of instances of human surgeons suturing. The surgeon can only reach a difficult kidney tumor because the robot's data yield precise coordinates from an MRI that make it recognizable amid occlusions. When an error is made, the identification of a unified and identifiable source of the error is not possible. The figure who would be required to give a clear-sighted account is an impossible figure.

Chapter Outlines

In sum, my cloud ethics has a principal concern with the algorithm's double political foreclosure: the condensing of multiple potentials to a single output that appears as a resolution of political duress; and the actual preemptive closure of political claims based on data attributes that seek recognizability in advance. Confronted with this double foreclosure, each chapter of this book elaborates a speculative strategy for reinstating the partial, contingent, and incomplete character of all algorithmic forms of calculation. In their partial and incomplete way of generating worlds, we can locate their ethicopolitics. Part I, "Condensation," comprises two chapters that detail how algorithms condense and reduce the teeming multiplicity of the world to a precise output. Chapter 1, "The Cloud Chambers," examines how algorithms are acting through cloud data architectures to produce a new paradigmatic alliance between sovereign authority and scientific knowledge. Developing an analogy with the scientific experiments of the cloud chamber of early twentieth-century particle physics, where the chamber made it possible to perceive otherwise invisible subatomic particles, I explore the capacities of the cloud in cloud computing. How does the cloud apparatus render things perceptible in the world? I address the character of cloud architectures across two distinct paradigms. The first, Cloud I, or a spatiality of *cloud forms*, is concerned with the territorial identification and location of data centers where the cloud is thought to materialize. Here the cloud is understood within a particular history of observation, one where the apparently abstract and obscure world can be brought into vision and rendered intelligible.

This notion of cloud forms, I propose, has led to a distinct ethicopolitical emphasis on rendering algorithms explainable or legible. Cloud I is founded on a misunderstanding of the nature of algorithmic reason within the cloud, so that the cloud is thought to obscure or obfuscate what is "really" going on in the world. On the contrary, I propose, the cloud is not an obfuscation at all but is a means of arranging the models for otherwise incalculable processes for a condensed decision. In my second variant—Cloud II, or the spatiality of *a cloud analytic*—the cloud is a bundle of experimental algorithmic techniques acting on the threshold of perceptibility. Like the cloud chamber of the twentieth century, contemporary cloud computing is concerned with condensing that which cannot be seen, rendering things perceptible and actionable.

Chapter 2, "The Learning Machines," turns to the question of machine learning algorithms and their complex and intimate relationships with what we think of as human practices. Through a study of the deep neural network algorithms animating surgical robots, I show the entangled composite bodies of surgeons, robots, images of organs, and cloud-based medical data of past surgeries—all of which learn together how to recognize and to act amid uncertainty. Where there has been public concern and moral panic around machine learning, what is most commonly thought to be at stake is the degree of autonomy afforded to machines versus humans as a locus of decision. I propose that the principal problem resides not with machines breaching the limit but in sustaining a limit point of the autonomous human subject—the oft-cited "human in the loop"—who is the locus of decision, agency, control, and ethics. Such an autonomous human disavows the *we* implicated in the entangled learning of humans with algorithms. Much of this entangled learning takes place in a space of play between a target output and an actual output of the algorithm. The indeterminate combinations of weights, parameters, and layers in the algorithm are the traces of rejected pathways and alternative correlations.

Part 2, "Attribution," elaborates the practices of attribution through which algorithms write themselves into being in the world. Taking seriously the attribute as it is understood in computer science, the chapters connect this attribution back to genealogies of writing and ethics. In chapter 3, "The Uncertain Author," I am concerned with how the search for ethical codes to govern algorithms has located an author function in the "source code" of algorithms. When an algorithm appears to have precipitated a crisis or to have caused a harm, often the reflex response is to seek out its origin: Who designed the model? Who wrote the code? Who or what labeled the training data? Was it biased? Who is the author of the algorithm? In this chapter I explain the limits of locating an ethicopolitical response in the authorship of source code. Focusing on the algorithmic techniques for natural language processing—where a corpus of literary texts are used to train an algorithm to recognize style and sensibility—I suggest that the algorithm's ways of being in the world are not all present in the source code and, indeed, substantially exceed the design of an authoring subject. The authorship of the algorithm is multiple, continually edited, modified, and rewritten through the algorithm's engagement with the world. Juxtaposing novelists' accounts of their own uncertain authorship with computer scientists' reflections on how the fragments of an algorithm come together in acts of writing, I suggest the impossibility of identifying a definitive author.

To invoke the call for attributing authorship is not only insufficient as critique, but it also risks amplifying the ability of the algorithm to bind together a unity of incompatible, fraught elements as though the difficulties and differences could be resolved. In place of the call for securing responsibility via authorship, I propose that there is a possibility of ethicopolitics in the act of writing, in the bringing together of scattered elements, and in the opening of a space of uncertainty. A cloud ethics must begin to reopen the act of writing as a site of ethical significance. The profound uncertainty that is brought to the writing of an algorithm—"I do not know how adjusting this weight is changing the output"; "I cannot be sure how my training data has produced these clusters"; "I am curious whether shifting this threshold will reduce the false positives"—invites an iterative process of writing that is never completed. Here is an ethicopolitical tension that is worth holding on to—the algorithm promises to complete everything, to condense to a single optimized output and action, and yet it enacts a process of writing that opens on to an indeterminate future.

Chapter 4, "The Madness of Algorithms," addresses the moments when it appears that an algorithm has acted in a state of frenzy or has departed from its otherwise rational logic. From the lethal accidents of autonomous vehicles during tests, to the racist hate and misogyny of the Twitter chatbot Tay, an ethical frame often seems required to somehow rein in the worst excesses or to restore reasonableness to autonomous actions. Yet, philosophy has long grappled with the problem of madness and, specifically, how the identification of madness shores up and sustains the domain of reason. Reflecting on the conjoined histories of ideas of reason and madness, I propose that one cannot speak of the madness of the algorithm except in relation to the very form of reason the algorithm embodies. While the contemporary moral panic at each moment of the madness of algorithms urges us to police ever more vigilantly the line between reasonable and unreasonable actions, understood as a threshold, this line is precisely the condition of possibility of algorithmic rationality. Algorithms cannot be controlled via a limit point at the threshold of madness because the essence of their logic is to generate that threshold, to adapt and to modulate it over time. In short, my argument is that when algorithms appear to cross a threshold into madness, they do, in fact, exhibit significant qualities of their mode of reasoning. Understood in this way, the appearance of a moment of madness is a valuable instant for an ethicopolitics of algorithms because this is a moment when algorithms give accounts of themselves.

Contra the notion that transparency and the opening of the black box secure the good behavior of algorithms, the opacity and clouded action exhibited in the excesses and frenzies of algorithms have a different kind of fidelity to the account. Throughout this book I argue that, when viewed from the specific propositional arrangements of the algorithm, particular actions that might appear as errors or aberrations are in fact integral to the algorithm's form of being and intrinsic to its experimental and generative capacities. I am advocating that we think of algorithms as capable of generating unspeakable things precisely because they are geared to profit from uncertainty, or to output something that had not been spoken or anticipated. Of course, this is not a less troubling situation than the one in which some controls are sought on the worst excesses of the algorithm. On the contrary, it is all the more political, and all the more difficult, because that which could never be controlled—change, wagers, impulses, inference, intuition—becomes integral to the mode of reasoning.

Part 3, "Ethics," develops a set of tactical routes for a cloud ethics to follow, each of these insisting on a different kind of weightiness from the calculative weights of adjustable probabilities. The apparent lightness of an optimized single output is afforded the full weight of undecidability and the difficulty of decision. In chapter 5, "The Doubtful Algorithm," I suggest that doubt, and more precisely the idea of doubtfulness, can be a tactical point of intervention for a cloud ethics. The overwhelming logic of algorithmic systems in our society is that they can optimize decisions made where there is profound doubt and uncertainty. Indeed, the algorithm is offered as a means of deploying doubt productively so that, for example, the doubt-ridden voter or doubtful consumer can be clustered as "having a propensity to be influenced" and can be targeted with personalized media. A specific form of truth telling, established in the "ground truth" of the data environment from which the algorithm learns, has asserted its dominance in the governing of societies with algorithms.

By contrast, and cutting against the grain of the dominance of definiteness as algorithms act on doubt, I seek to reinstate doubtfulness as what N. Katherine Hayles calls "embodied actuality" within the calculative architecture of the algorithm.[68] Though at the point of optimized output, the algorithm places action beyond doubt, there are multiple branching points, weights, and parameters in the arrangements of decision trees and random forest algorithms, branching points at which doubt flourishes and proliferates. A cloud ethics reopens the contingencies of this multiplicity, giving life to the fallible things that the algorithm has learned about the world, rendering the output indelibly incomplete.

In each of the chapters, I maintain a faithfulness to the specificities of how particular algorithms learn via, and generate worlds through, their relations with data. I am wary of "algorithm talk" when it is asserted generally and without specificity, for different algorithms are as varied in their logics and grammars as languages are, and these differences, as Adrian Mackenzie argues, should be made to matter.[69] Thus, for example, in discussions of facial recognition systems, I would want to know what kinds of CNNs are being used as the basis for recognizing a face from data inputs. Though each of the chapters draws out the arrangements of propositions of particular algorithms and how they recognize, misrecognize, and target through their relations with other algorithms, data, and humans, I have also foregrounded the fallibility of the algorithm, its incompleteness and contingency. In short, a specific algorithm will always exceed its name, its type, its genus, for it is immanently modifying itself through the world. In chapter 6, "The Unattributable," I address more directly those critics who would ask what use my cloud ethics would be to their campaigns or movements against the injustices of specific algorithms. Following a lecture on machine learning that I gave in Copenhagen, a lawyer in the audience asked what my cloud ethics could ever look like in law. I had similar questions from lawyers at the Turing Institute in London. My response, in short and in those moments, was to say that it would be a crowded court in the sense that my approach multiplies the possible sites of intervention and responsibility. This notion of the crowded space, or forum, has stayed with me throughout the research and the writing of this book. What kind of political claim could be brought in the name of a cloud ethics? How does one forge a form of responsibility for the future onward life of something like an attribute as it becomes attached to others? What is the power of the unattributable as the set of qualities that cannot be attributed to a subject?

Chapter 6 maps out what it means to be together ethicopolitically—to be associated as a society, a community, a movement, a gathering of protesters— as an association of partial associations or attributes. The chapter concludes the book along three lines of argument for a cloud ethics: apertures, opacity, and the unattributable. When machine learning algorithms segment a social scene, generating clusters of data with similar propensities, everything must be attributed. Yet, that which is unattributable does remain within the scene, exceeding the algorithm's capacity to show and tell, as well as opening onto a different kind of community and a different mode of being together, being ethicopolitical.

Part 1

Condensation

1

The Cloud Chambers
Condensed Data and Correlative Reason

What I wouldn't give now for a map of the ever constant ineffable?
To possess, as it were, an atlas of clouds.
  —David Mitchell, *Cloud Atlas*

A Beautiful Sight

In physicist Charles Thomson Rees Wilson's Nobel lecture of 1927, he described the cloud chamber experiments he had conducted since the late nineteenth century, and how they had transformed the capacities of modern physics into the twentieth century. From the observatory at the summit of Ben Nevis, Wilson had witnessed what he depicted as "the wonderful optical phenomena" of the formation of clouds.[1] Inspired by what he had observed, Wilson spoke of cloud formations that "greatly excited" his interest so that he "wished to imitate them in the laboratory."[2] For Wilson, the capacity to reproduce in science the formation of clouds in nature became the means to advance understanding of the condensation physics of meteorology, and with it the taxonomy and classification of cloud forms. In Wilson's laboratory, the scientific practice of knowing clouds followed a path from observation to mimetic representation to classification.

When Wilson began to experiment with the formation of clouds in his cloud chamber apparatus, however, what he discovered was an unanticipated potential to see something not otherwise perceptible; phenomena that exceed the paradigms of observation and classification. In contrast to the telescopes of the observatory, where the optic instruments had brought objects into a line of human sight, Wilson's cloud chamber became a different kind of apparatus, one that brought something into perceptibility that could not otherwise be seen. Though ionizing particles, such as alpha, beta, and gamma radiation, could not be observed directly, the condensation trail in Wilson's cloud chamber showed the particle's trajectory of movement. Recalling his experiments with supersaturation levels, temperature, and the expansion of gas in his chambers, Wilson reflects in his lecture, "I came across something which promised to be of more interest than the optical phenomena which I had intended to study." His cloud chamber experiments afforded him a "means of making visible and counting certain individual molecules or atoms which were in some exceptional condition." Though Wilson had set out to reproduce mimetically the formation of clouds in nature, in fact his experiments generated traces of the electrically charged atoms and ions previously unavailable to the sciences.[3]

What Wilson's cloud chamber ultimately made possible for the twentieth century's study of particle physics was the ability to photograph and to perceive the movement of particles in an exceptional state (figures 1.1 and 1.2). As historian of science Peter Galison writes in his account of Wilson's work, "After the cloud chamber the subatomic world was suddenly rendered visualizable."[4] Charged or ionized particles could not be observed directly with optic devices, as with the instruments of microscopy or telescopy, but their traces and trajectories of motion appeared indirectly via the cloud tracks condensing on the nuclei. As Wilson reflects on his 1911 experiments, "I was delighted to see the cloud chamber filled with little wisps and threads of clouds" so that "the very beautiful sight of the clouds condensed along the tracks of the alpha particles was seen for the first time."[5] The cloud chamber apparatus, conceived for the human *observation* of processes of formation in nature, had become a technique for rendering *perceptible* the movement of objects beyond the thresholds of human observation.

Almost exactly one century on from the publication of Wilson's first cloud chamber images, the idea of *the cloud* is once more describing the advent of processes at scales that appear to transcend the observational paradigm and exceed our human capacities to see, to know, and to understand. Indeed, the *cloud* in cloud computing is widely held to derive from the mapping of infrastructures of computer networks, where the visualization of a figurative cloud stands in for the vast complexity of the internet.[6] In the twenty-first century, cloud computing promises to have effects on the very episteme of our world, analogous to the effects of the cloud chamber on what could be rendered knowable in twentieth-century physics.

More precisely, the advent of cloud computing opens space for a renewed

Figure 1.1
Charles Thomson
Rees Wilson's cloud
chamber expansion
apparatus (1912–13).
Science Museum
Group Collection,
Board of Trustees of
the Science Museum,
London.

twinning of science and technologies of perception with forms of political sovereignty. Such renewal signals an extension of historical technologies of imaging, mapping, and intelligence data collection into new algorithmic modes of analysis and data correlation. In February 2015, for example, seventeen US intelligence agencies—including the Department of Homeland Security (DHS), National Security Agency (NSA), Central Intelligence Agency (CIA), Department of the Navy, Department of the Army, National Geospatial Intelligence Agency, Defense Intelligence Agency, Office of the Director of National Intelligence (ODNI), Department of the Air Force, Federal Bureau of Investigation (FBI), State Department, and Drug Enforcement Agency (DEA)—launched the ICITE program for the cloud-based storage, sharing, and analysis of intelligence data. Here, once more, one can find the promise of a "beautiful sight" celebrated by Wilson, the making of pictures otherwise unavailable to the senses. ICITE (pronounced "eyesight") is the Intelligence Community Information Technology Enterprise (figure 1.3), a $600 million cloud-computing

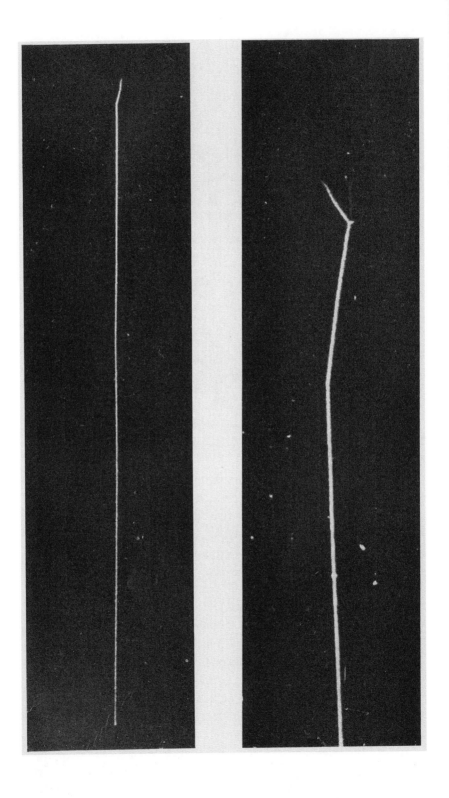

contract with Amazon Web Services (AWS) and Cloudera, providing a new intelligence and security data infrastructure. ICITE, it is promised, will "allow agencies to share data much more easily and avoid the kind of intelligence gaps that preceded the September 11, 2001, terrorist attacks."[7] In this specific sense, the data geographies of the cloud can be read as a response to the 9/11 Commission findings of a failure to analyze across data *silos* held by different agencies.[8] As the former US director of national intelligence James Clapper explained at the launch of the ICITE program, cloud computing allows government authorities to "discover, analyze and share critical information in an era of seemingly infinite data."[9] The CIA's chief intelligence officer, Douglas Wolfe, similarly expressed his hopes that the government security agencies would get "speed and scale out of the cloud, so that we can maximize automated security."[10] Let us reflect on the terms once more: *discover, analyze, infinite data, speed and scale out of the cloud, maximize automated security.* The cloud promises to transform not only what kinds of data can be stored, where, and by whom, but most significantly, what can be generated and analyzed in the world. In short, the cloud form of computation is not merely supplying the spatial accommodation of large volumes of data in server farms, but offers the means to map and to make perceptible the geography of our world in particular ways.

The architecture of the cloud is defined spatially by the relations between algorithms and data. As the cloud becomes ever more closely intertwined with geopolitical decisions—from sharing and acting on intelligence data, to border controls, urban policing, immigration decisions, and drone strikes—then what is the precise nature of these algorithmic practices of data gathering, analyzing, and knowing? In this chapter I address the political character of cloud computing across two distinct paradigms. The first, which I term Cloud I, or *cloud forms*, is concerned with the spatiality of data stored in data centers and analyzed in cloud architectures. This Cloud I paradigm sustains an ontology of observation, representation, and classification, and it encloses ethicopolitics within this logic. In the second mode, Cloud II, or a *cloud analytic*, I propose that the computational regimes of the cloud transform what or who can be rendered perceptible and calculable. In contrast with Cloud I's linear logics of observation, representation, and classification, the generative logics of Cloud II are engaged in perception, recognition, and attribution. As the scientific his-

Figure 1.2 Cloud chamber tracks of the alpha rays of radium, among the earliest of C. T. R. Wilson's cloud chamber images. Proceedings of the Royal Society, London.

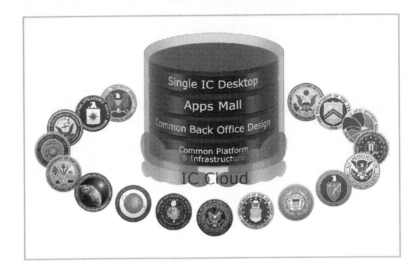

Figure 1.3   US Intelligence Community Information Technology Enterprise (ICITE) architecture and its imagination of a "world of possibilities." Screenshot archived in 2014, when the Amazon Web Services contract to build ICITE was first announced.

tory of the cloud chamber is concerned with "the character of an instrument and the effects produced with it," as Svetlana Alpers has put it, I am interested here in understanding the character of the instruments of cloud computing and their generative effects.[11]

Cloud I: "Geography Matters in the Cloud"

Cloud I is concerned with the identification and spatial location of the data centers where the cloud is thought to materialize. Indeed, as computer science began to document the emergence of cloud computing, the idea of geography came to have a specific meaning defined by where data and programs are spatially located. In a 2008 Association of Computing Machines (ACM) forum devoted to the advent of cloud computing, a transformation is described "in the geography of computation," with "data and programs" being "liberated" as they are "swept up from desktop PCs and corporate servers and installed in the compute cloud."[12] Such accounts of the cloud appeal to a geography of "scalable" computation, which is thought to change radically with the rise of big data and the concomitant need for flexible storage and computational power.[13] There is need for some caution, however, in understanding the geography of the cloud primarily in relation to the rise of twenty-first-century volumes of digital data. Indeed, the emergence of cloud computing has important origins in grid computing, distributed scientific data, and, perhaps most significantly, the notion of computing as a public utility. "Computing may someday be organized as a public utility," speculated computer scientist John McCarthy in his MIT lecture of 1961, so that "the computer utility could become a new and important industry."[14]

In the second half of the twentieth century, the imagination of computing as a scalable public utility emerged. By 2006, when AWS launched its Elastic Compute Cloud (EC2), the architecture of cloud computing had begun to develop the three components now most widely recognizable as the cloud: *Infrastructure as a service*, in which hardware, servers, storage, cooling, and energy are supplied; *platform as a service*, in which the software stack is accessed via the cloud; and the *applications layer*, in which data analytics capacity—or the deployment of algorithms—is supplied via the cloud. Across the components of cloud architectures, the emphasis is on scalable computing, where the client pays for what they have used, combined with distributed computing, where multiple concurrent users can share and combine their data, their algorithms, and their analyses.

Understood as an architecture where data and algorithms meet and entangle, cloud computing's notion of a single computer hosting multiple simulated or virtual machines becomes actualized in space in a particular way.[15] In this respect, the whereabouts of simulated machines is not unknown at all, but rather the cloud is actualized in data centers, located in places within economies of land, tax rates, energy, water for cooling, and proximity to the main trunks of the network. As Benjamin Bratton writes, the "cloud is not virtual; it is physical even if it is not always on the ground, even when it is deep underground."[16] Hence, within the vocabulary of computer scientists at least, "geography" is said to "matter in the cloud."[17] In this limited sense of geographies of the cloud, the architecture of data centers has come to count as a spatial and geographic phenomenon in a way that the spatialities of the algorithm have not. When computer scientists ask "where is the cloud," what they denote as "geographical questions" concern the data centers thought to "underlie the clouds," their "physical location, resources, and jurisdiction."[18] When, for example, Google locates a new data center in the tax-friendly state of Georgia, or the Swedish internet service provider Bahnhof installs a data center in the cool confines of a former nuclear bunker under Stockholm, or Sun Microsystems

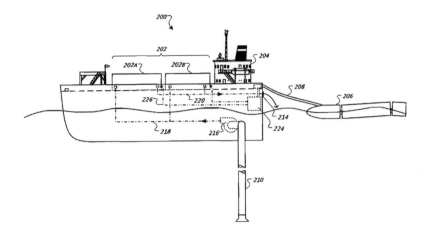

Figure 1.4   The Google-patented "water-based data center," hosting a floating platform of servers, powered and cooled by seawater. US Patent Office US752520782.

designs a portable data center inside a standard shipping container, the matter of geography is thought to reside in the spatial location of data storage.[19] Of course, these materializations of the cloud have ethicopolitical significance, so that when Google patented the floating data center (figure 1.4), the offshore potentials of cloud techniques beyond jurisdiction were actualized in these vast data ships.

Let us capture this geography of the spaces of cloud computing as Cloud I. This imagination of the cloud as a dispersed yet spatially located array of data centers is present in computer science, and that has extended into geographic and even political and geopolitical debate. So, for example, following the Snowden disclosures of the extent of US authorities' access to European citizens' data via US data centers, the EU sought initially to develop a "European cloud," in which they might store safely European data under European jurisdiction.[20] Similarly, following the US subpoena and mining of European financial transactions, the major Society for Worldwide Interbank Financial Telecommunications moved its cloud-computing provision to an underground data center in Switzerland, and the Canadian government has legislated for what it calls "data sovereignty," meaning domestic public data traffic must not leave Canadian territory. Understood as a spatial and territorial question of the jurisdiction of data technologies, then, the political response has been to seek to render the cloud intelligible and governable.

The unintelligibility of the cloud as a form of governing or a locus of geopolitical power has profoundly political implications. Following the exposure of the NSA's PRISM program in 2013, for example, the UK Intelligence and Security Committee (ISC) of Parliament—the sole body responsible for public oversight of security and intelligence powers in the UK—called the then foreign secretary, Philip Hammond, to give evidence before the committee. At the time of his evidence, Hammond was the final signatory on all warrants authorizing the interception and analysis of external communications data, conventionally understood as where one end of the communication is located externally to the UK. In his testimony, this figure of final sovereign authority manifestly failed to understand the complex spatial modalities of data stored, transferred, or analyzed in the cloud:

Q: The distinction between internal and external to the UK is important because there are tighter restrictions on analysing internal data. . . . But if the sender and recipient of an email are both in the UK, will it be treated as internal even if the data is routed overseas on its journey?

A: So, I think . . . er . . . and I invite my colleagues to step in if I get this technically wrong. . . . But I think . . . er . . . it's an internal communication. [At this point, the civil servants flanking the minister lean in to advise.]

Let me finish my train of thought. . . . [M]y understanding is, er, because of the technical nature of the internet . . . it is possible it could be routed to servers outside the UK. . . . Please correct me if I misinterpreted that. . . . I'm sorry, I have misled you in my use of terms. . . . I'm trying to be helpful.

Q: Well[,] you will be relieved to know that was the easy one. Now, the case of social media . . . if all of my restricted group of Facebook friends are in the UK . . . and I post something to Facebook, surely that should be internal?

A: [Following whispers from civil servants] Erm, . . . no[,] actually[,] if you put something on Facebook and the server is outside of the UK it will be treated as an external communication.

Q: What about cloud storage, where no other person is involved at all. It may be my decision to upload photographs to Dropbox. Would these communications be regarded as external because they are on US servers?

A: Aaah . . . er. My colleagues will . . . oh . . . well. . . . Yes[,] I am advised if the server is overseas they will be regarded as external.[21]

The 2014 UK foreign secretary's testimony before the ISC exposes the difficulties and limit points of a territorialized juridical form in the face of cloud computing. In Cloud I, where the geography of cloud forms is everything, the cloud has become centered on *where* data is collected and stored. Indeed, the Anglo-American juridical tradition has founded its privacy protections precisely on the consent required for lawful storage and collection. Hammond's evidence before the committee, however, exemplifies how it may be the very unintelligibility of cloud forms that enables and facilitates contemporary surveillance and intelligence gathering. As Kate Crawford and Jason Schultz argue, the new predictive "approaches to policing and intelligence may be both qualitatively and quantitatively different from surveillance approaches," and thus enable "discriminatory practices that circumvent current regulations."[22] Yet, Crawford and Schultz go on to suggest that an alternative space for democratic oversight might lie in what they call "a right to procedural data due process," where constraints and oversight mechanisms are placed on the algorithmic processes of data analysis.[23] Even as the cloud overflows and exceeds the categories and practices of bureaucracy and law, what has come to be at stake ethicopolitically has become a struggle to wrest the cloud back into a form over which one can have oversight, to expose its "bias" and demand neutrality, to make it comprehensible and accountable in democratic fora, and to render the cloud bureaucratically and juridically intelligible.

Among the critical geographic and artistic accounts of cloud computing, the desire to wrest the cloud into an intelligible form similarly finds expression in methods of visualization. The geographer and artist Trevor Paglen seeks to "make the invisible visible," reflecting that "the cloud is a metaphor that obfuscates and obscures" the material geographies of the "surveillance state."[24] Paglen's work is concerned with bringing the geopolitics of cloud computing back into a human line of sight through visualization. His methods deploy optical devices of many kinds to bring back into human vision that which would otherwise exceed the limits of observation. His ghostly images of the NSA's data centers are photographs taken at night with a long-focus lens from a helicopter (figure 1.5); and his photographs of the secret installations of military and drone bases in the Nevada desert are taken with adapted telescopic instruments of astronomy.[25]

The optical instruments deployed by Paglen belong to a paradigm of observation in which, as Peter Galison describes, one is offered "a direct view" of things otherwise "subvisible."[26] As Paglen accounts for his own work: "My intention is to *expand the visual vocabulary* we use to see the US intelligence community. Although the organizing logic of our nation's surveillance appa-

Figure 1.5   NSA headquarters, Fort Meade, Maryland. Trevor Paglen, 2014.

ratus is invisibility and secrecy, its operations occupy the physical world. Digital surveillance programs require concrete data centers; intelligence agencies are based in real buildings. . . . [I]f we *look in the right places* at the right times, we can begin to glimpse the vast intelligence infrastructure."[27] So, for Paglen the challenge is to "expand the visual vocabulary" to see more clearly the actions of algorithms, and to bring into vision the things that would otherwise be obfuscated by the cloud. Similarly, for the artist James Bridle, "one way of interrogating the cloud is to look where its shadow falls," to "investigate the sites of data centres" and to "see what they tell us about the real disposition of power."[28]

Yet, what are the "right places" and "right times" to look and to observe an apparatus? How does one investigate something describable as the real disposition of power? Indeed, what ways of seeing would be appropriate to what art historian Jonathan Crary calls a "relocation of vision" taking place with computation, or the visual vocabulary appropriate to the digital mediation of cultural objects and urban scenes identified by Gillian Rose and Shannon Mattern?[29] If the cloud is to be observed and revealed in the secret glimmering buildings of the NSA's data centers in Paglen's images, then could his "real buildings" also be located in other places? Could they be observed, for example, in the rented North London offices where a small team of physics graduates write algorithms for risk-based security?[30] Must the material geography of cloud computing be found in the buildings or territories where it is thought to actualize? Could the "right place" to look also be in the spatial arrangements of an experimental clustering algorithm used in anomaly detection, or in the generative logics of emergent machine learning algorithms?[31]

My point is that the desire to "open the black box" of cloud computing and to expand the visual vocabulary of the cloud, to envision the cloud and its properties in geographic space, dwells within and alongside the paradigm of observation. In Stephen Graham's work on cities and warfare, for example, he writes of "systems of technological vision" in which "computer code tracks and identifies."[32] Critical scholarly responses to the cloud—through imaginaries of verticality and the stack—have thus emphasized overwhelmingly spaces of sight and oversight, where observational power is foregrounded. Crucial aspects of these technologies, however, do not operate on the terrain of human vision and, indeed, harness perceptual power on a horizontal threshold of connection and correlation.

And so, Cloud I poses questions within an observational mode: Where is it?; What type is it?; Can we map it?; Can we recognize it? As with the early classification of cloud forms, when Luke Howard first proposed names for cirrus, cumulus, and stratus in 1803, a linear system of genera and species was proposed to "enable people to think coherently about clouds" (figure 1.6).[33] The system of cloud form classification was later described as "quite ridiculous for clouds," because they are not fixed forms but ever in formation, and indeed analog algorithms were devised to diagram the observational pathways of cloud formation. In short, Cloud I sustains the idea that one can have a more beautiful sight, a means of seeing more clearly and rendering something coherent and intelligible. The telescope and camera Paglen brings to the scene of data deployment belongs to a particular history of observation, one that Donna Haraway describes as "visualizing technologies without apparent limit," like "the god trick of seeing everything from nowhere."[34]

What might it mean for us to commit instead *not* to enable coherent thinking about the cloud? If one determines instead to "stay with the trouble," as Haraway has put it, of partial and indeterminate lines of sight, then all apparently coherent technologies of observation become what Haraway calls "active perceptual systems" with "partial ways of organizing worlds."[35] In the second variant I discuss here—Cloud II, drawing on Peter Galison's distinction between mimetic and analytical forms of scientific instruments—cloud computing appears as a *cloud analytic*.[36] Here, the cloud is a bundle of experimental

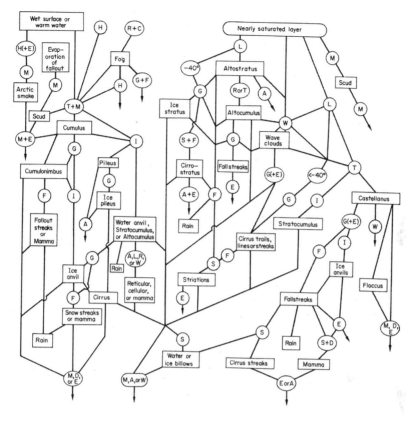

Figure 1.6   Analog algorithm for the classification of cloud mechanisms, 1967. Scorer and Wexler, *Cloud Studies in Colour*.

algorithmic techniques acting in and through data to modify the threshold of perceptibility itself. Put simply, as analytic the cloud resides not within a paradigm of observation, representation, and classification but instead within perception, recognition, and attribution. As Galison reminds us, in the cloud chamber "we do not actually see things," though what we perceive "has a close relation to them," what he calls an "almost seeing" of otherwise subvisible entities.[37] Understood thus, to say the cloud somehow obfuscates a real world of politics is to miss the point somewhat. The cloud is not an obfuscation; far from it.[38] Like the cloud chamber of the twentieth century, contemporary cloud computing is about rendering perceptible and actionable (almost seeing) that which would otherwise be beyond the threshold of human vision. Where some claim the cloud makes the geographies of power in our world unintelligible, I propose that it is becoming an important element of what Karen Barad calls the "condition of intelligibility" of our world.[39]

Cloud II: "Changing the Aperture of Observation"

What I call Cloud II, or the cloud as analytical gathering of algorithms with data, displaces the question, Where is the cloud?, replacing it with, How does the cloud render the world perceptible and analyzable? In this way, Cloud II witnesses the extension of algorithmic modes of reason that surface something for perception and action amid an otherwise profoundly uncertain environment. As historian of science Paul Erickson and his colleagues have traced meticulously in the historical emergence of algorithmic rationality, the profound uncertainties of the Cold War nurtured a desire to extend mathematical logic into the realm of decision. The decision procedures and cybernetic methods of algorithm appeared to extend the faculties of human reason so that they "no longer discriminated among humans, animals, and machines" in the capacity to analyze, to decide, and to act.[40] What we see here is the entwining of human and machine modes of reasoning such that what Henri Bergson calls the "organs of perception" of the world are composite beings formed through the relations among humans, algorithms, data, and other forms of life.[41]

Understood in terms of the intertwined faculties of human and machine, the contemporary spaces of cloud computing exceed the territorial geographies of the location of data centers, becoming instead a novel political space of calculative reasoning. Returning to the site of the ICITE program, what kinds of perceptions and calculations of the world become possible with the algorithmic instruments that gather in cloud space? When the seventeen US intelligence agencies upload or analyze data in ICITE, they access software as a service, so that they are not merely "joining the dots of their data," as the post-9/11 measures claimed, but are in fact combining their modes of analysis. Among the available ICITE platforms is Digital Reasoning software, a set of machine learning algorithms for analyzing and deriving meaning from data across government intelligence databases and open source social media data streams. Describing the Synthesys application available to analysts via ICITE, Digital Reasoning claims that its algorithms are able to "read," "resolve," and "reason" from the correlations in cloud data. The algorithms are said to "extract value from complex and opaque data" and to "determine what's important" among the "people, places, organizations, events and facts being discussed," ultimately "figuring out what the final picture means" by comparing it with "the opportunities and anomalies you are looking for."[42] What this means in terms of the practices of the algorithms is that they identify clusters within the data (to read), derive attributes from those clusters to make them recognizable into the future (to resolve), and compare the output of the algorithms with the target output of the analyst (to reason). This is an iterative and experimental process in which the humans and machines feel their way toward solutions and resolutions to otherwise indelibly political situations and events.

Though Digital Reasoning's machine learning algorithms are now the mainstay of US government defense analysis, they were developed initially for anomaly detection in the wholesale financial industry. The frenzy of the 2008 financial crisis included trading strategies that were in breach of regulations, leading to significant losses for institutions. In the aftermath of the crisis, machine learning algorithms were developed to analyze "terabytes of emails per day to detect hints of insider trading."[43] The cognitive computing software performs the role of what N. Katherine Hayles calls a "cognizer," carrying out the cognitive functions to generate norms and anomalies from the patterns in vast datasets and, as it does so, deciding what or who will come to matter amid the occlusions.[44] At the level of the algorithm, there is profound indifference to the context of whether these norms and anomalies pertain to financial trades or the movement of insurgent forces—what matters is precisely the capacity to generate a final output that can be acted on. By 2015, when Digital Reasoning was pitching to government security and defense officials in Washington, DC, the CEO Tim Estes was offering the machine learning tools to "sift through sensor data, emails and social media chatter," bringing "structure to human language" and "changing the aperture of observation."[45]

What would it mean for algorithms to change the aperture of observation? Let me agree, curiously and peculiarly, with this vendor of cloud-based algorithms to the DHS and the NSA and say, yes, indeed, the aperture of observation is changing, though not in such a way that the promised complete final picture is delivered to the analyst. Indeed, the technology of the aperture is concerned with a specific and discrete opening onto a scene, one in which the notion of a completeness beyond the aperture is occluded and screened out. The aperture, from the Latin *aperire*, to open or to uncover, is not primarily an optic but is rather a means of opening onto or uncovering the world. The point at which a machine learning algorithm condenses the output of multiple layers to a single output is also an aperture in the sense that it is an opening or uncovering of attributes and relations that would not otherwise be perceptible. Put simply, with Cloud II, where we are interested in the analytic, what matters is not so much seeing the "where" of the data as it is the capacity to extract patterns and features from data to open onto targets of opportunity, commercial and governmental.

With the advent of cloud computing, the aperture of observation becomes an aperture of "almost seeing," in Peter Galison's terms, or a means of "correlating and synthesizing large volumes of disparate data" so that action can take place based on what is almost seen of the world.[46] As one analyst puts the problem, "It allows us to say correlation is enough. We can stop looking for models, throw the data at the biggest computing clusters the world has ever seen and let algorithms find the patterns."[47] So, *correlation is enough*. Let us reflect on this claim of an adequacy of correlative associations. In the pages that follow, I propose three characteristics of correlative cloud reason, and I suggest why this form of algorithmic reason has significance for our ethicopolitical present.

Condensing Traces

The algorithmic techniques of Cloud II involve *condensing traces*, practices not primarily concerned with seeing or bringing into vision, but rather focused on engaging a subvisible world by inferring from the traces and trajectories that condense at indeterminate points. Returning to my analogy with the apparatus of the cloud chamber, by the mid-twentieth century, when Charles Wilson's cloud chamber was being used in subatomic physics, the "purpose" of the instrument was described as being "to study the motion of ionizing particles from records of the drops condensed on ions formed along the trajectories followed by these particles."[48] The motion of particles could not be observed directly, but their trajectory could be perceived obliquely, via the visible drops condensed on the ions—the cloud "tracks." Figure 1.7 shows one of the best-known cloud chamber photographs, C. T. R. Wilson's image of alpha-emitting thorium, with the cloud originating from an alpha ray passing through the chamber, its "trajectory disturbed in two places," as recorded in the atlas of cloud chamber images.[49] The newly available images of radioactivity made the object perceptible via the records of condensed drops on the ions, observing the motion of the particles obliquely. In the compelling images from the cloud chamber, one can locate a capacity to perceive the movement of otherwise subvisible entities. The *chamber* of the cloud chamber is akin to an apparatus, in Michel Foucault's terms, in that it "inserts the phenomena in question" within a "series of probable events."[50] In this sense, to condense a trace is not to show or to reveal something existing in the past, but rather to establish a condensed series of possible correlations between entities and events. The cloud chamber apparatus is experimental to the extent that it is concerned with probable tendencies and trajectories, condensing a larger volume down to the probable event.

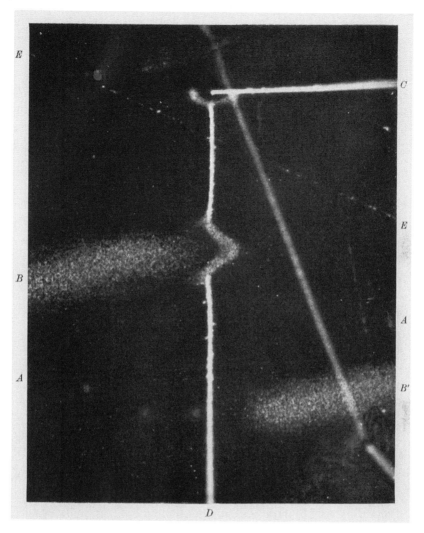

Figure 1.7   Cloud chamber tracks of alpha-emitting thorium. C. T. R. Wilson, c. 1923, *Proceedings of the Cambridge Philosophical Society.*

Like the cloud chamber apparatus of the twentieth century, contemporary cloud computing is a form of experimental chamber or apparatus. It condenses and reduces the volume of data down to that which is probable or possible in the algorithmic model. When Digital Reasoning claims that their algorithms determine what is important in war or in finance, this is because their machine learning apparatus is generating what matters in a series of probable events. As Karen Barad writes on the nature of scientific apparatuses, they are boundary-making practices that decide "what matters and what is excluded from mattering," as they "enact agential cuts that produce determinate boundaries and properties of 'entities' within phenomena."[51] In condensing the probable data traces of what matters in the world, cloud computing enacts the matter of the person of interest at the border, the possible future disruptive protest event in urban space, the acts of fraud or insider trading, or the chains of association of no-fly lists, blacklists, and kill lists—like beaded drops of condensed data making action possible.

Though the movement of an entity cannot be observed directly in the chamber, it is perceived obliquely through tracks and trajectories of mobility. Indeed, Cloud II as a mode of computational power poses significant questions for political and geographic accounts of what it might mean to "secure the volume" or the "volumetric," or to have a "politics of verticality."[52] The analytical techniques available in the cloud do not strictly act on the earth from some novel spatial dimension "above" or "below" the ground, but rather enroll the space of calculation itself. Cloud computing acts on the vast volume of data traces through a series of algorithmic apertures that condense and distill the thing of interest. In contrast with a securing *of* the volume, the pursuit of security *through* the volume precisely reduces and condenses the volume by means of the correlations within the data. The so-called cognitive-computing applications in the ICITE cloud, for example, use pattern recognition and sentiment analysis to identify political protests, civil unrest, and "atypical" gatherings or events. Cognitive computing renders perceptible to the analyst "what matters" in the political scene, using the volume of cloud-based digital data precisely to reduce and flatten the field of vision. The relation between volume and flatness thus becomes one in which the tracks of association and correlation enact the horizon of possibility for the analyst. The volume is radically condensed down to the target data elements, like beaded drops on ionizing particles through which future trajectories of motion can be inferred.

Discovering Patterns

The algorithms gathering in Cloud II involve the *discovery of patterns*, which is a highly specific calculative device deployed in a volume of data. The repository of data in the US intelligence community's cloud, for example, is described as a "data lake" in which the "same raw data" can be analyzed with "statistical functions," such as conventional regression, and with "machine learning algorithms" of "data discovery."[53] Here, the relation of the data lake to cloud computing is metaphorically understood as the formation of clouds from the water vapor rising from lakes into the atmosphere.[54] While the application of statistical analysis to intelligence data involves the analyst beginning with a deductive query, or hypothesis, and building rules to test that hypothesis in the data, the advent of cloud computing presents the analyst with a volume and variety of data too great for conventional human hypothesis or deduction. In the context of a security paradigm that seeks out the uncertain possible future threat, the volume of data in the lake—much of it transactions and social media data—is analyzed with machine learning techniques that promise to yield previously unseen patterns via processes of "knowledge discovery."

In contrast to a deductive form of reasoning by hypothesis testing, machine leaning algorithms deploy abductive reasoning, so that what one will ask of the data is a product of the patterns and clusters derived from that data. As Luciana Parisi writes on the algorithmic logic of abduction, "algorithms do not simply govern the procedural logics of computers" but take "generative forms driven by open-ended rules" or "derive rules from contingencies."[55] Understood in these terms, the machine learning algorithms deployed in the contemporary intelligence cloud are generative and experimental; they work to identify possible links, associations, and inferences. Such abductive forms deploy a distinct kind of causal reasoning, different from deductive reasoning, in which "deductions support their conclusions in such a way that the conclusions must be true, given the premises," and closer to "fallible inferences," in which "the possibility of error remains."[56] Put simply, in the cloud analytic of Cloud II, the algorithms modified through the patterns in data decide, at least in part, which fallible inferences to surface on the screen of intelligence analyst, drone pilot, or border guard.

The rise of correlative forms of reasoning in machine learning has serious implications for the ethicopolitics of algorithms, not least because error, failure, or fallibility are no longer conceived as problems with a model; rather, they have become essential to the model's capacity to recognize abnormalities and generate norms. Abductive models observe the output signal of the algorithm and adjust the layers of probability weighting to optimize the fit between the input data and the output signal. For this reason, programs such as ICITE claim to be "uncovering answers to questions the analysts don't already know," where the question is not a hypothesis but a speculative experiment.[57] As one of the designers of t-digest pattern-detection algorithms explains, in describing the modification of a model in response to data inputs, "As small fluctuations occur in the event stream, our model can adjust its view of normal accordingly."[58] What this means is that the data inputs and the algorithm mutually modify to optimize the output. The norms and anomalies of the model are thus entirely contingent on a whole series of adjustments, including the adjustment of weights, parameters, and the threshold to tolerate particular error rates. "You can move the slider for a lower false positive rate if that is what you want," I am told by one designer of facial recognition algorithms, "but that will always increase the false negative rate, so you need to decide what your tolerance is."[59] As the politics of our world becomes understood as an "event stream" filtered through infinitely adjustable apertures and thresholds, the apparatus of the cloud deploys algorithms, such as t-digest, to generate its malleable view of what is normal in the world from its exposure to the data stream. To adjust the view of normal, to move a threshold, or to decide a tolerance—these are the ethicopolitical dispositions of algorithms as they act on the world.

Returning to my analogy with the cloud chamber apparatus, the twentieth-century physicists were also engaged in abductive forms of reasoning and pattern detection that exceeded the deductive testing of hypothesis. The cloud chamber made it possible to detect previously unseen and unknown particles via the patterns of unusual or abnormal cloud tracks. "The central problem of the interpretation" of cloud chamber "exploratory photographs," as described in the physicists' guide to cloud chamber technique, was the "recognition of the particles involved in a particular event." To interpret the detected patterns of scattering, cascades, and showers, the physicists inferred from the attributes, or the "characteristic features of particle behaviour."[60] They would not begin with a hypothesis and test it in the chamber, for the uncertainties and contingencies of particle behavior had become the focus of their inquiry. The cloud chamber played a crucial role in the identification of hitherto unknown subatomic particles, rendered detectable through the generation of surprising new cloud tracks and trajectories. In the text accompanying the famous Rochester atlas of cloud images, Nobel physicist Patrick Blackett writes: "The last two decades have seen an increasing use of two experimental methods, the cloud chamber and the photographic emulsion, by which the tracks of sub-atomic particles can be studied. All but one of the now known unstable elementary particles have been discovered by these techniques. . . . This involves the ability to recognise quickly many different sub-atomic events. Only when all known events can be recognised will the hitherto unknown be detected."[61]

The atlases of "typical" cloud chamber images definitively did not offer the scientist a taxonomy or classificatory system for identifying particles, forming the rules of form for unknown future particles. Rather, the images provide a kind of training dataset, allowing the scientists to become sensitive to the patterns and clusters of cloud tracks, so that they could recognize the disturbances and fluctuations of a new event. The discoveries of new particle behaviors—the first glimpse of the muon or the positron in the cloud chamber, for example—were not strictly observations of an unknown object, but more precisely perceptions of something in close relation to it: the patterns involved in an event. As Peter Galison reminds us, the cloud chamber images "traveled" and were "widely exchanged, stored and reanalyzed by groups far distant from the original photographic site."[62] In this sense, *the cloud chamber is the site*, just as *the cloud analytic is the site* in cloud computing, through which the event is recognized via its patterns, and where the algorithm and the analyst are trained in the art of recognition.[63]

Archiving the Future

The recognition of traces in Cloud II involves an archiving of the future, in which particular future connections are condensed from the volume of the data stream and rendered calculable. In a sense, the algorithms of Cloud II are relatively indifferent to the past as a series of data points or significant events. What matters most to Cloud II is the capacity to generate a particular future of dispositions, propensities, and attributes. When AWS supplies cloud computing to corporations and governments, the applications layer is configured as an "app store" so that users can select the analytics tools they want, paying for what they use. As one UK government client of AWS explained to me, "We are cloud first" because "we want platforms; we want to plug in to different suppliers."[64] This flexible deployment of cloud data was an important element of AWS's tender for the ICITE program, with James Clapper, announcing, "We have made great strides, the applications are in the apps mall, and the data is in the cloud."[65] In fact, of course, the significance is that the algorithms and the data dwell together in AWS cloud space, opening the possibility for seemingly infinite calculability, or what Hayles calls "infinitely tractable" data.[66] The aperture opens onto a world of optimized targets, threats, and opportunities so that the analyst experiences a sense of reach into possible futures. The security, policing, or intelligence analyst is thus reimagined by the state to be a desiring and wanting consumer, with the "apps mall and stores available from the desktop," selling to users "thousands of mission applications" and resembling "what Apple provides through iTunes."[67]

Let us reflect for a moment on what the NSA or CIA analyst browsing the ICITE apps mall might find to assist them in their missions. Among the thousands of applications, Recorded Future offers natural language processing and sentiment analysis algorithms to "scrape the web" for signals of possible future threat: "We constantly scan public web sources. From these open sources, we identify text references to entities and events. Then we detect time periods: when the events are predicted to occur. *You can explore the past, present and predicted future of almost anything in a matter of seconds.* Our analysis tools facilitate deep investigation to better understand complex relationships, resulting in actionable insights."[68] Consider the claim: the data stream of social media contains all the attributes of incipient future events, so that one can explore the past, present, and predicted future of almost anything. Recorded Future's applications run their algorithms across the boundary of public and private cloud computing so that the analyst can explore the correlations between, for example, classified structured data in the DHS's files and the language and sentiment analysis of so-called open source Twitter feeds and Facebook posts. In this way, the technology enables action in the present, based on possible correlations between past data archives (such as national security lists) and archives of the predictive future.

With this instrument of Cloud II, the analytic is everything. Archived data in Recorded Future becomes unmoored and derivative of its context, even the so-called dirty, or noisy, data no longer muddying the calculation but rendered useful. As the security analysts draw together social media "junk" data with other labeled entities, tagging metadata and sharing with other agencies, diverse data elements are rendered commensurate and made actionable geopolitically. As Orit Halpern suggests in her account of how algorithmic rationality became a governmental and social virtue, digital computation changes the nature of the archive. The nineteenth-century form of "static" archiving and repository is supplemented in the twentieth century by what Halpern calls "an active site for the execution of operations."[69] Similarly, in the nineteenth-century cloud chamber's attempt to reproduce nature, the cloud tracks had been considered spurious dirt effects, not for scientific archiving. Yet, it was the tracks that became the thing of interest, the atlas of cloud chamber images knowing only the event of the track itself. Contemporary cloud computing is an active site for the execution of operations, as understood by Halpern, in which the archive is generative of particular imagined futures.

The archivization of specific data elements with applications such as Recorded Future, then, produces particular futures; "the archivization produces as much as it records the event," as Jacques Derrida writes.[70] As the photographic recording of the cloud tracks within the cloud chamber archived the possibility of recognizing future subatomic particles, so the digital recording of social media data in cloud computing archives the possibility for future actions at the border or on the city street. Understood in this way, the spatial power of the data center as "archive" could be critically challenged, as we see in Trevor Paglen's images, while leaving entirely intact Halpern's "active site of operations," a site capable of acting indifferent to the "where" and the "what" of data.

And yet, there are creative practices of resistance within Cloud II that offer an alternative sensing of the archive as an active and generative site. In James Bridle's installation *Five Eyes* (figures 1.8 and 1.9)—commissioned by the Victoria and Albert Museum (V&A) for their public archive-focused exhibition *All of This Belongs to You*—the artist invites the viewer to consider anew the relations between the archive and potential futures. Bridle passed the V&A's 1.4 million digital object records through an intelligence analysis system. The algorithms "extract names, things and places, and create searchable connections between seemingly disparate objects," with the resulting connections "difficult to grasp, often inscrutable to the human eye, reflecting the mechanical calculus that was used to generate them."[71] Displayed in a series of five glass cabinets, juxtaposed with the woven artifacts of the V&A's tapestry galleries, the objects surfaced for our attention by the algorithms are placed atop a "stack" of analog museum files. The object displayed is thus generated in and through the archive, through the intimate connections learned by the algorithms. One cannot meaningfully trace how the object displayed has become the thing of interest, or why it is the optimized output from the actions of the algorithms. In Bridle's rendering of the archive, one can sense the "upheaval in archival technology" noted by Jacques Derrida, the infrastructure making a claim on the future, on the infrastructure of the "archivable event."[72]

The Cloud Atlas

When a group of particle physicists showed me their cloud chamber experiments, I had expected the thing of interest around which we would gather would be the cloud tracks—the wispy trajectories of particles that had so captivated Charles Wilson. Instead we gathered around the apparatus, the physicists animated by much discussion on the optimal point of cooling, and whether thorium is a useful radioactive element for the experiment. One of

Figures 1.8 and 1.9   James Bridle's *Five Eyes* installation, displayed in the Victoria and Albert Museum's tapestry galleries. Author photographs.

the group had worked at CERN with the large hadron collider, commenting that "there is no reason why we couldn't have discovered the Higgs Boson using a cloud chamber, but it would take an inordinately long time."[73] So, for the scientists, something exists with potential to be detected—manifest in the alpha tracks and cosmics in the chamber—but this potential entity is rendered perceptible by the specific and contingent arrangement of an experimental apparatus. With the smallest of adjustments to the arrangement of the chamber, the physicists dramatically changed the conditions for what could be traced of the particle's trajectory.

The gathering of scientists and one curious geographer around the experimental apparatus of the cloud chamber resonated with other kinds of gatherings and disputes I have observed around an experimental apparatus. When the designers of algorithms for the detection of credit card fraud gathered around their experimental model, they were concerned to optimize the perceptibility of an anomaly. "The problem with rules," they explain, is that "novel fraud patterns will not be detected."[74] To detect something novel, they suggested, the rules of an algorithm must be generated through the iterations of output signals to input data. Their machine learning algorithms for detecting emergent forms of fraud, then, are not experimental in the sense of not yet validated, but are specifically "experimental" in their capacity to adjust parameters and bring something novel into existence. Isabelle Stengers describes the "paradoxical mode of existence" of subatomic particles in that they are simultaneously "constructed by physics" and yet "exceed the time frame of human knowledge," so "the neutrino exists simultaneously and inseparably 'in itself' and 'for us,' a participant in countless events."[75] In the lines that follow here, I conclude the chapter by commenting on why this matters for our contemporary moment, when the specific apparatus of cloud computing brings something into being, discovering associations and relations otherwise unknowable.

Amid the many contemporary calls to bring algorithms into vision, or to establish human oversight, it is imperative that we formulate accounts of the ethicopolitics of algorithms that do not replay the observational paradigm of Cloud I or the classificatory forms related to that paradigm. In Cloud I, vision is the sovereign sense, afforded both the apparent objectivity of the "most reliable of senses" and the means of securing the state's claim to sovereign violence.[76] And yet, the machine learning algorithms operating in programs such as ICITE work to render perceptible that which could never be observed directly, that which could not be brought into view as with an optical device. The algorithms for identifying insider trading, for credit scoring, or for generating intelligence data from social media bring trajectories and thresholds into being. They are generative and experimental techniques capable of perceiving a thing of interest without seeing it as such. The algorithms available on state analysts' version of iTunes—with their promises of digital reasoning and recorded futures—generate a target of interest through subvisible experimentation. The Cloud II apparatus is not geared to a deductive science of *observation*, *representation*, and *classification*, but instead signals a paradigm of *perception*, *recognition*, and *attribution*.

The algorithms identifying the attributes of people and things, then, do not merely observe or record historical data on past behaviors or events. If they did observe and classify behaviors, then perhaps it might be an adequate ethical response to insist on the transparency of the processes. With machine learning programs such as ICITE, a person or entity of interest emerges from the correlations and patterns of condensed data: financial transactions, travel patterns, known associates, social media images, and affects derived from sentiment analysis. The data archive of the neural net algorithms of Cloud II is, as Orit Halpern proposes, "indeterminate in terms of the past," so that it is not possible to identify how the present calculation was arrived at.[77] Little pieces of past patterns enter a training dataset and teach the algorithm new things; the designer of the algorithm experiments with thresholds to optimize the output; new people and things enter a validation dataset to further refine the algorithm; and on and on iteratively, recursively making future worlds. And let us not forget that with this correlative reasoning, sovereign decisions are made: to stop this person at the border, to detain this group as they travel on the subway to a downtown protest, to target this vehicle as it approaches a checkpoint, or to approve or deny this asylum claim.

Isabelle Stengers has written that scientific experiments work through "the power to confer on things the power of conferring on the experimenter the power to speak in their name."[78] Understood in this way, whether an experiment can be said to have worked or to yield proof is of lesser significance than whether it confers the power to speak or to make a claim. In one sense, Wilson's cloud chamber experiments failed in that they did not primarily advance the understanding of the formation and classification of cloud forms. A greater power, though, was afforded to the particle physicists who were able to speak in the name of things that would otherwise exceed their observation. As algorithms written for casino or credit card fraud travel to border control or to security threat analysis, I propose that cloud computing similarly confers on algorithms the power to confer on the analyst the power to speak in their name. What do they say when they speak? What kinds of claims do they make?: here are the people and things with a link to terrorism; here are the possible fraudulent asylum claims; here are the optimal targets for the next drone strike; here are the civil uprisings that will threaten the state next week. The claims that are spoken in cloud computing programs, such as ICITE, confront our fallible, intractable, fraught political world with a curious kind of infallibility. In the cloud, the promise is that everything can be rendered tractable, all political difficulty and uncertainty nonetheless actionable. The ICITE app store marketplace available on the screens of analysts renders the politics of our world infinitely reworkable—the "geopolitical events" in the correlative calculus, a kind of geopolitical cloud chamber. As Timothy Cavendish, a protagonist in David Mitchell's novel *Cloud Atlas*, muses, "What I wouldn't give now for a map of the ever constant ineffable? To possess, as it were, an atlas of clouds."[79] Programs such as ICITE make just such a dangerous promise in the algorithmic governing of society—a kind of atlas of clouds for the ineffable, a condensed trace of the trajectories of our future lives with one another.

The Learning Machines
Neural Networks and Regimes of Recognition

The peculiarity of men and animals is that they have the power of adjusting themselves to almost all the features [of their environment]. The feature to which adjustment is made on a particular occasion is the one the man is attending to and he attends to what he is interested in. His interests are determined by his appetites, desires, drives, instincts—all the things that together make up his "springs of action." If we want to construct a machine which will vary its attention to things in its environment so that it will sometimes adjust itself to one and sometimes to another, it would seem to be necessary to equip the machine with something corresponding to a set of appetites.
　　—Richard Braithwaite, "Can Automatic Calculating Machines
　　　Be Said to Think?"

Springs of Action

In January 1952 the BBC recorded what was to be the world's first public debate on the mathematics and ethics of machine learning. Participating in the discussion of the question "Can Automatic Calculating Machines Be Said to Think?" were Manchester mathematicians Alan Turing and Max Newman; their neurosurgeon colleague Geoffrey Jefferson; and Cambridge moral philosopher Richard Braithwaite. Turing explains that he has "made some experiments in teaching a machine to do some simple operation," but that "the machine learnt so slowly that it needed a great deal of teaching." Jefferson is skeptical of the use of the verb "to learn" in relation to the machine, and he interjects to ask Turing a question: "But who was learning, you or the machine?"[1]

This distinction, between humans and machines as the locus of learning is of great significance to the neurosurgeon, for whom the electronic circuits of computing machines were not analogous to the "fragments of the nervous system" he encountered in the fleshy materiality of the human brain.[2] Yet, Jefferson's question prompts Turing to reflect on his own embodied experience of experimenting with his machines. In the audio recording of the debate, Turing can be heard to pause for a moment's reflection before responding: "I am inclined to believe that when one has taught it [the machine] to do certain things[,] one will find that some other things one had planned to teach it are happening without any special teaching being required." In response to Jefferson's question of who was learning, the mathematician or the machine, Turing responds, "I suppose *we both* were."[3]

The entangled "we both" of mathematician and machine, learning together, expresses Turing's belief that intuition is a mathematical faculty that "consists of making spontaneous judgements which are not the result of conscious trains of reasoning." For Turing, the intuitive faculty is entangled with what he calls the "ingenuity" of the building of rules as arrangements of propositions.[4] The iterative relationship between intuition and ingenuity in mathematical reasoning necessarily entangles the mathematician's affective and haptic relations to a puzzle with the making of a formal axiom or logic. The human and machinic elements of mathematical learning, then, are not so readily disaggregated for Turing. Though for the philosopher in the discussion, Richard Braithwaite, it is the unique "peculiarity of men and animals" that they are able to learn intuitively by "adjusting themselves to almost all the features of their environment," his notion of a "spring of action" afforded by appetites nonetheless calls to mind today's capacities for machine learning algorithms to learn and to generate things in excess of their taught rules.[5] The 1950s radio discussion of the character of machine learning did, in some respects, envisage a future world in which machines would exceed the rules-based decision procedure and extend to the affective pull of intuitions and appetites for data.

Thus, even in the mid-twentieth century, the mathematics and philosophy of machine learning was centered on the entangled relations of humans and machines. The question, as articulated in the debate on automatic calculating machines, was "Who was learning, you or the machine?," and Turing's reply was "We both were." In this chapter, I focus on the *we* invoked by Turing in this public debate precisely because it runs against the grain of contemporary moral panics amid machine autonomy and algorithmic decisions that appear to be beyond the control of the human. On the contrary, the *we* of machine learning is a composite figure in which humans learn collaboratively with algorithms, and algorithms with other algorithms, so that no meaningful outside to the algorithm, no meaningfully unified locus of control, can be found. In contemporary machine learning, humans are lodged within algorithms, and algorithms within humans, so that the ethicopolitical questions are concerned less with asserting human control over algorithms and more with how features are extracted and recognized from a teeming data environment.[6] In short, the ethicopolitics of machine learning algorithms is located within the figure of the *we*—in the very relations to ourselves and to others implied in the *we* who have a spring of action.

In our contemporary moment, the "we both were" extends the already multiple body via the sinewy and invasive techniques of deep learning and neural network (neural net) algorithms.[7] The extended *we* of the multiplicity of data to which the learning algorithm is exposed heralds an intimate communion of the learning machines with a vast and incalculable *we*: all of us, all our data points, all the patterns and attributes that are not quite possessed by us.

In the pages that follow, I begin by taking up the theme of intuitive learning via the extracted features of a data environment in the context of the twenty-first-century advent of surgical robotics. At the level of this specific type of deep neural network algorithm, there is no technical distinction between learning actions for robot surgery and learning actions for robot weaponry. Across different domains of life, these algorithms are concerned with translating the input data from their environment into a "feature space," mapping the features into clusters of significance, and extracting the object of interest for the action.[8] Thus, algorithms designed to save lives, via robot surgery, or to end lives, via robot warfare, share the same arrangements of propositions. In following the machine learning of surgical robotics, I am concerned to capture the impossibility of establishing definitive boundaries of good and evil in relation to algorithms. The machine learning algorithms deployed in robot surgery do save lives through the lower infection rates of noninvasive methods, but they also endanger life through error and miscalculation. My point is that the ethicopolitics of machine learning algorithms cannot be mapped onto the parameters of good and evil or the securing and imperiling of life. With the extraction of feature spaces, machine learning algorithms are actively generating new forms of life, new forms of boundary making, and novel orientations of self to self, self to other. To begin with robot surgery is to begin in a place where one could never definitively draw a line delineating the algorithmic moral good from some sense of immorality or evil.

Intuitive Surgery: Making the Singular Cut

The intuitive relation to mathematics noted by Turing finds a contemporary form in robotic surgical systems such as Intuitive Surgical's da Vinci robot. The application programming interface (API) and the cloud storage architecture of the da Vinci robot contain the data residue of multiple past human and machine movements. Figure 2.1 displays the "surgical gestures" of the movements of surgeons' hands on the remote console, as mediated through the robotic instruments of the da Vinci.[9] The surgical procedure modeled here is a routine four-stitch surgical suture to close an incision. Though the surgical gestures involved in the cutting and stitching of flesh are a matter of haptic routine for human surgeons, for the designers of the algorithms, the objective is to model the optimal suturing motion so that future human *and* robot surgeons have their intuitive movement shaped by the ingenuity of the model. As the Johns Hopkins computer scientists building the model explain, the process begins with the "automatic recognition of elementary motion" from the extraction of features in the data environment. The model extracts seventy-eight features, or "motion variables," from the vast quantity of video and sensor data archived by the da Vinci robot (figures 2.2 and 2.3)—twenty-five feature vectors from the surgeon's console, and fourteen from the surgical instruments attached to the patient-side robotic arms.[10] In the surgical gesture, the movement of the human hand is thus thoroughly entangled with the remote console and the surgical scalpel held by the robot's hand. The scientists describe the juxtaposed map of surgical gestures: "The left [top] plot is that of an expert surgeon, while the right [bottom] is of a less experienced surgeon."[11] Here, algorithms are enrolled to recognize surgical gestures and to extract the features of movement, to actively distribute cognition across human surgeons and robots, and to optimize the spatial trajectory of the act of suturing flesh. The future surgeon will learn to suture flesh optimally, via the robot's simulation functions, and the robot surgeon will learn to suture autonomously from the data of past gestures of expert human surgeons. Though the computer scientists do envisage autonomous actions by the robot—with "the possibility to automate portions of tasks, to assist the surgeon by reducing the cognitive workload"—this apparent autonomy is entirely contingent on the layered learning from models of past entanglements of human and robot gestures.[12]

The spring to action of surgical machine learning is not an action that can be definitively located in the body of human or machine but is lodged within a more adaptive form of collaborative cognitive learning.[13] Intimately bound together by machine learning algorithms acting on a cloud database of medical

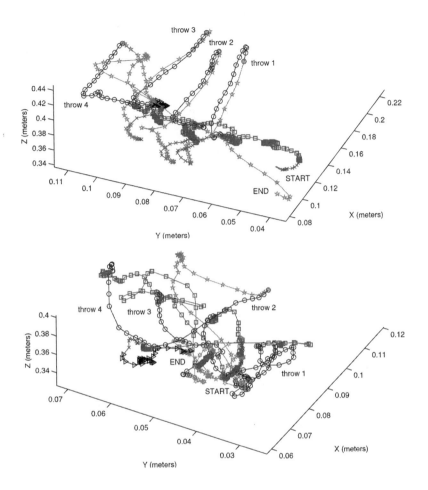

Figure 2.1   A plot of the trajectories of surgical gestures during a four-stitch suturing procedure. The top plot shows the trajectories of an expert surgeon, while the bottom plot shows the trajectories of an inexperienced surgeon, as extracted and analyzed by the surgical robot. Lin et al., "Automatic Detection."

data, the *we* of surgeon and robot restlessly seeks an optimal spring of action—the optimal incision, the optimal target of tumor or diseased organ, the optimal trajectory of movement. Intuitive Surgical's robots hold the promise of the augmented vision and precise trajectories of movement of a composite being of surgeon and machine. As one of the UC Berkeley robotics scientists describes what they term "iterative learning," machine learning allows "robotic surgical assistants to execute trajectories with superhuman performance in terms of speed and smoothness."[14] The drive for superhuman learning iterates back

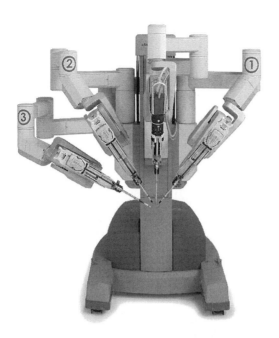

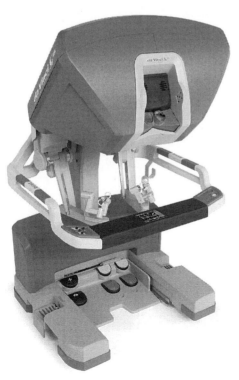

Figures 2.2 and 2.3
The da Vinci surgical
robot. Intuitive
Surgical, 2018.

and forth across multiple gestures. As the hand and eye movements of a singular surgeon are tracked in the da Vinci's API, they commune with a mathematical model generated from a vast multiplicity of data traces of past surgical gestures. These gestures, in turn, modify the learning model to optimize and augment the future trajectories of future surgeons and robots as yet unknown. As the philosophers and scientists in the 1952 debate anticipated, the machine learning algorithms have something close to appetites, extracting and modeling features from the plenitude of cloud data, rendering springs of action, and acting on future states of being.

During the course of following the surgeons and robots of one world-leading oncology department in a UK teaching hospital, I began to note the many occasions when a surgeon used "we" in place of "I" to describe their daily collaborations with surgical robots. This *we* who learns is expansive. It includes the many humans in the research group and the surgical team, but also the many human and technical components of the simulation of a surgery—the multiple layers of medical imaging, video data from past surgeries, and algorithmic models that together composed a kind of virtual presurgery.[15] For example, when the UK surgical team were preparing to conduct a new surgical procedure on a specific type of tumor, they collaborated with other US surgeons who had previous experience of the procedure. This was not merely a dialogue between human experts, however; the research team also imported the data from the US surgeries, inputting them in an algorithmic model and experimenting with the parameters for a new context. As Rachel Prentice documents in her meticulous ethnography of surgical education, "surgical action must be made explicit for computers" so that "bodies and their relations in surgery are reconstructed" in a form that "can be computed."[16] This reciprocal and iterative relationship between human surgeon and computer is what she calls a "mutual articulation" in which "bodies affect and are affected by" one another.[17] When the movement of a surgeon's hands is rendered computable in the robotic model, Prentice suggests that this "instrumentalization" overlooks the "tacit" and "tactile experience" of surgery, such as the "elasticity of a uterus or the delicacy of an ovary."[18]

In Prentice's reading of surgical technologies, the tacit, tactile, and intuitive faculties of surgery define the human as the locus of care and embodied judgment and decision. Yet, in giving their accounts of working with robots, I found that human surgeons testify to their own body's capacities coming into being in new ways. What it means to be intuitive, to touch or to feel an organ, for example, alters with the advent of machine learning modalities of surgery. Seated at their virtual environment console, the surgeons access video feed

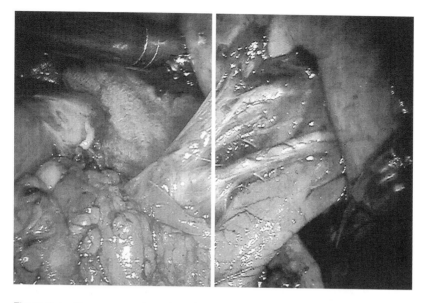

Figure 2.4   Fluorescence imaging of renal parenchyma, as viewed from the surgeon's console. Intuitive Surgical, 2018.

images from the endoscopic arm of the robot (figure 2.4). These images are overlaid with MRI and the fluorescence images of tumors so that, as one of the da Vinci computer scientists explains, "the system provides over a thousand frames per second and filters each image to eliminate background noise."[19] The work of the algorithms here is to extract the features of interest for perception by the surgeon, surfacing the optimal image from a background noise of teeming data.

One surgeon described to me the daily work of obstetric surgery with her da Vinci robot, noting that through the expanded mediated space, she is able to "see the unseeable" and "reach the unreachable" within the patient's body.[20] In contrast to Prentice's sense that human touch—and with it, judgment and decision—is evacuated and instrumentalized by surgical robotics, here the entangled touch of surgeon and robot invokes different relations of judgment and decision. The surgeon's relations to herself and to others—to her patients past and present, her operating theater colleagues, the robots, images, tumors, surgical instruments—are altered in and through the machine learning algorithms. "Touch engages us in a felt sense of causality," writes Karen Barad, so that "touch moves and affects what it effects."[21] Where Prentice foresees a loss of responsibility as the human surgeon's touch is evacuated by the robot, I propose instead an extension of responsibility to that which extends and exceeds human sensibility. The difference is an important one. The terrain of uncertainty regarding whether a specific tumor is operable without major damage to surrounding organs, for example, shifts with the 360-degree mobility of the robot's wrist—what can be precisely touched, sensed, and extracted from the body is altered. Indeed, leading computer scientist and artist Ken Goldberg, alongside his carefully presented accounts of the development of stochastic models and neural networks for the performance of surgical excision by robots, writes essays on the insensible and uncanny worlds that open up with his algorithms. Jochum and Goldberg's account of the "experiential uncanny" describes how robot actions "stretch the boundaries between the animate and inanimate" in new directions, serving to "challenge our beliefs about what, precisely, separates humans from machines."[22] The computer scientist's reflections on the embodied and intuitive capacities of his algorithms—to change the nature of what can be seen, reached, touched, or learned—run against the grain of a surgeon's hands being rendered computable by autonomous machines. Instead, machine learning algorithms work with the incomputable to open up new worlds of intuitive and insensible action.

As machine learning algorithms engage in "stretching the boundaries," the object that is surfaced for perception and action communes intimately with data on the events of past surgeries. This communion on *what is optimal*—the cut, the incision, the surgical strike—belongs properly to a composite being within the cloud analytic I describe in the previous chapter. The da Vinci data are no longer territorially limited to the memory of a specific robot, a server, an individual surgeon, or a group of scientists. Rather, the machine learning algorithms are deployed in a cloud architecture that yields the data residue of many millions of past surgeries. Lodged inside the actions of the singular cut—itself bordered by algorithms optimizing the thresholds of the instrument's trajectory—are the multiple data fragments of other entangled composites of surgeon, software developer, programmer, neural network, patient's body, images, and so on. The singular cut, within which teems a multiplicity, is present also in the autonomous vehicle, the drone, the smart borders system—always also with multiple data fragments lodged within.[23] In every singular action of an apparently autonomous system, then, resides a multiplicity of human and algorithmic judgments, assumptions, thresholds, and probabilities.

The Impossible Figure of the "Human in the Loop"

The neural network's capacity to learn by extracting features from its data environment has made it flourish in the algorithmic architectures of drones, autonomous vehicles, surgical and production robotics, and at the biometric border.[24] This capacity to learn something in excess of taught rules has also characterized the public concern and ethical debates around autonomous systems. Whether in the neural net algorithms animating surgical robots, autonomous weapons systems, predictive policing, or cloud-based intelligence gathering, what is most commonly thought to be at stake politically and ethically is the degree of autonomy afforded to machines versus humans as a locus of decision. I suggest, however, that the principal ethicopolitical problem does not arise from machines breaching the imagined limits of human control but emerges instead from a machine learning that generates new limits and thresholds of what it means to be human. As legal cases proliferate amid the errors, when the spring of action happens at the point of surgical incision, smart border, or drone strike, they consistently seek out an identifiable reasoning human subject to call to account: a particular named surgeon, a specific border guard, an intelligence analyst—the "human in the loop."

In Hayles's field-defining book, *How We Became Posthuman*, she proposes that the "distributed cognition of the posthuman" has the effect of complicating "individual agency."[25] Hayles does not argue that a historically stable category of human has given way, under the forces of technoscience, to an unstable and disembodied posthuman form. On the contrary, the conception of the human and human agency was, and is always, a fragile and contingent thing. As Hayles writes,

> The posthuman does not really mean the end of humanity. It signals instead the end of a certain conception of the human, a conception that may have applied, at best, to that fraction of humanity who have had the wealth, power, and leisure to conceptualize themselves as autonomous beings exercising their will through individual agency and choice. What is lethal is not the posthuman as such but the grafting of the posthuman onto a liberal humanist view of the self.[26]

Hayles's concerns for the grafting of the posthuman onto the figure of an autonomous liberal subject echo across the making of intuitive machine learning worlds. Though technologies such as Intuitive Surgical's robot actively distribute and extend the parameters of sight, touch, and cognition into posthuman composite forms, their ethical orientation is defined solely in relation to the control of an autonomous human subject. While human surgeons speak of an indeterminate *we* who learns, decides, and acts, nonetheless the capacity for judgment retains its Kantian location in the unified thought of a reasoning human subject.[27] Thus, when a violence is perpetuated or a harm is registered—damage, prejudicial judgment, or death—the only ethical recourse is to an imagined unified entity who secures all representations. So, for example, in a series of legal cases against Intuitive Surgical, the reported harms include the rupture of tissue, burns, and other damage to organs, severed blood vessels and nerves, loss of organ function, and fatalities.[28] In these juridical cases, where the robot's machine learning algorithms fail to recognize or to grasp precisely the outline of the target, what is sought is a unified locus of responsibility—a company, a negligent surgeon, or a hospital—an entity imagined juridically to be autonomous and unified, whose choices and agency can be held to account. Similarly, when autonomous weapons systems make errors in their target selection, or cause "collateral damage" amid the so-called precision strike, the ethical appeal is made to an accountable "human in the loop" of the lethality decision.[29] The notion of an ethical decision thus appears in the form of a reasoning human subject or a legal entity with a capacity to be a first person *I* who is responsible.

Yet, where would one locate the account of a first-person subject amid the limitless feedback loops and back propagation of the machine learning algorithms of Intuitive Surgical's robots? When the neural networks animating autonomous weapons systems thrive on the multiplicity of training data from human associations and past human actions, who precisely is the figure of *the* human in the loop? The human with a definite article, *the* human, stands in for a more plural and indefinite life, where humans who are already multiple generate emergent effects in communion with algorithms.[30] Recalling Geoffrey Jefferson's 1952 question, "Who was learning, you or the machine?," and Turing's reply, "We both were," the human in the loop is an impossible subject who cannot come before an indeterminate and multiple *we*.

Perhaps what is necessary is not a relocated human ethics—of feedback loops and kill switch control—for a world of the composite actions of human and algorithm. What is necessary, I propose, is an ethics that does not seek the grounds of a unified *I* but that can dwell uncertainly with the difficulty of a distributed and composite form of being. As machine learning changes the relations we have to ourselves and to others, the persistent problems of a Kantian unity of thought is newly dramatized by algorithmic formulations of learning and acting. To begin to address this different kind of ethicopolitics, one must dwell with the difficulty, as Donna Haraway suggests, making cloudy trouble for ourselves methodologically and philosophically.[31] Such a tracing of algorithmic threads as they meander through unilluminated space involves asking questions of how algorithms iteratively learn and compose with humans, data, and other algorithms. To be in the dark, to dwell there in an undecidable space, is to acknowledge that our contemporary condition is one in which the black box of the algorithm can never be definitively opened or rendered intelligible to reveal its inner workings. To trace the algorithm in the dark is not to halt at the limits of opacity or secrecy, but to make the limit as threshold the subject of study. Such a task begins by asking how machine learning algorithms learn things about the world, how they learn to extract features from their environment to recognize future entities and events, what they discard and retain in memory, what their orientation to the world is, and how they act. If intuition never was an entirely human faculty, and never meaningfully belonged to a unified *I* who thinks, then how does the extended intuition of machine learning feel its way toward solutions and actions? To this task I now turn.

Regimes of Recognition: How a Neural Network Makes the World

Allow me to begin by describing a scene—a laboratory designing machine learning algorithms for border and immigration control systems—where a series of neural networks learn to recognize people and things via the features in their data environments. One of the designers explains that his algorithms are trained on border and immigration data with many hundreds of thousands of parameters. He describes how he "plays with" his developing neural nets—taking the experimental model to the uniformed border operations team in the adjoining building to test it against the specific targets they are seeking in the algorithm's output.[32] This traveling of the model between laboratory and operations center consists of a series of questions about whether the algorithms are useful, or if they are "good enough." This question, Is it good enough?, illuminates some of the politically contested features of the algorithm's emergence. Though for the border operations team, "good enough" may be a measure of the algorithm's capacity to supply a risk-based target for a decision at the border, for the computer scientists, "good enough" means something quite different and specific. In computer science, a "good enough" solution is one that achieves some level of optimization in the relationship between a given target and the actual output of a model.[33] Understood in this way, it is not the accuracy of the algorithm that matters so much as sufficient proximity to a target. Put another way, the algorithm is good enough when it generates an output that makes an optimal decision possible. When the algorithm designers describe tuning or playing with the algorithm, they are experimenting with the proximity between the target value and the actual outputs from their model, adjusting the probability weightings in the algorithm's layers and observing how the actual risk flags generated by their model diverge or converge on the target.

The design of an algorithmic model, then, involves a contingent space of play and experimentation in the proximities and distances between the actual output and a target output. My concept of the space of play designates specifically the distance between a target output and an actual output of the model. This space of play, however, also opens onto an infinite array of combinatorial possibilities in terms of the malleable and adaptable inputs, parameters, and weights of the model. As one designer of machine learning algorithms for anomaly detection frames the question, "What is normal?" and "How far is far, if something is to be considered anomalous?"[34] Precisely this adaptive threshold between norm and anomaly was being negotiated between the laboratory and border operations. A small adjustment in the threshold will generate an entirely different set of outputs and, therefore, a change in the spring of action. I have observed this iterative process of playing with the threshold in multiple situations where algorithms are being trained for deployment, from police forces adjusting the sensitivity of a facial recognition algorithm to casinos moving the threshold for potential fraud:

> You must experiment to determine at what sensitivity you want your model to flag data as anomalous. If it is set too sensitively, random noise will get flagged and it will be essentially impossible to find anything useful beyond all the noise. Even if you've adjusted the sensitivity to a coarser resolution such that your model is automatically flagging actual outliers, you still have a choice to make about *the level of detection that is useful to you.* There are always trade-offs between finding everything that is out of the ordinary and getting alarms at a rate for which you can handle making a response.[35]

What is happening here is that the neural networks are learning to recognize what is normal and anomalous at each parse of the data. But, the shifting of the thresholds for that recognition embodies all the valuations, associations, prejudices, and accommodations involved in determining what is "useful" or "good enough." Sometimes, as with semisupervised machine learning, this regime of recognition emerges iteratively between humans, algorithms, and a labeled training dataset. In other instances, unsupervised machine learning will cluster the data with no preexisting labeled classifications of what is or is not useful or of interest. Even in this apparently unsupervised process, humans recalibrate and adjust the algorithm's performance against the target. In short, in all cases, machine learning algorithms embody a regime of recognition that identifies what or who matters to the event. Machine learning algorithms do not merely recognize people and things in the sense of identifying—faces, threats, vehicles, animals, languages—they actively generate recognizability as such, so that they decide what or who is recognizable as a target of interest in an occluded landscape. To adjust the threshold of what is "good enough" is to decide the register of what kinds of political claims can be made in the world, who or what can appear on the horizon, who or what can count ethicopolitically.

The kind of ethicopolitics I am opening up here is somewhat different from the attention that others have given to the inscription of racialized or other prejudicial profiles in the design of the algorithm.[36] Though identifying the human writing of prejudicial algorithms as a site of power is extraordinarily important, the regimes of recognition I have described actively exceed profiles written into the rules by a human. The machine learning algorithms I observed, from borders to surgery, and from facial recognition to fraud detection, are producing modes of recognition, valuation, and probabilistic decision weighting that are profoundly political and yet do not reside wholly in a recognizable human who writes the rules. They are also, of course, calculative spaces where prejudice and racial injustices can lodge and intensify, though not in a form that could be readily resolved with a politics of ethical design or the rewriting of the rules.[37]

The Hitherto Unseen: Detecting Figures, Detecting Objects

If machine learning algorithms are changing how something or someone comes to attention for action, then how does this regime of recognition come into being? Often the most apparently intuitive of human actions—to recognize a face in a crowd, to distinguish the features of a cat from a dog, to know how best to reach out and grasp an object—present some of the most difficult computational problems. When algorithms are understood as a series of programmable steps formulated as "if . . . and . . . then," it is precisely in the writing of the rules for the sequence that one decides the result: who or what will be of interest, or who or what can be recognized.[38] A common exemplar of the problem of recognizing the unseen is the capacity of algorithms to recognize handwritten digits.[39] The variability of the form of a figure—its "profile"—exposes the limit of rules-based algorithms that define the features in advance. How would one formulate rules for recognizing the handwritten number 3? In the traditional decision trees designed by J. R. Quinlan in the 1980s, the "product of learning is a piece of procedural knowledge that can assign a hitherto-unseen object to one of a specified number of classes."[40] To recognize an unseen object, the decision tree algorithm classifies it according to "a collection of attributes" describing its important features.[41] How would one begin to define the attributes of the figure 3 amid the variability of its form? As Quinlan acknowledged, the limits of the procedural knowledge of rules-based classifiers are encountered in unknown features that were absent in the training dataset. "The decision trees may classify an animal as both a monkey and a giraffe," wrote Quinlan, or the algorithm may "fail to classify it as anything."[42] Procedural classifiers such as decision trees, then, learn to recognize hitherto unseen objects according to the presence or absence of a set of properties encountered in the training data. Returning to the example of handwritten digits, would the variability of the form of the figure result in the decision that a 3 is not a 3? Or indeed that a 3 is not classifiable, or has the attributes of a 5? Though the recognition of handwritten figures is a common exemplar, the limit of procedural classifiers also applies to many other recognition problems, from facial recognition technologies to security threats, weather patterns, advertising opportunities, or the likely pattern of votes in an election. Put simply, while a rules-based classifier recognizes according to the profiled properties of an entity, the contemporary neural network algorithm learns to recognize via the infinite variability of features it encounters.

What does it mean to learn about the world in and through the variability of features in the environment? With the growing abundance of digital images and cloud data for training machine learning algorithms, the process of learning shifts from *recognition via classification rules* to *recognition via input data clusters*. Let us explore this further through the example of recognizing numeric figures. Figure 2.5 shows a simplified illustration of the spatial arrangement of a deep (multilayer) neural network—of the type I have sketched on many whiteboards at conferences and workshops. If the target of the algorithm is to optimize the likelihood of correctly identifying a handwritten digit, for example, then the training data will consist of a dataset of handwritten digits, each figure segmented to the level of pixels. What this means is that the algorithm does not learn to recognize the profile of the figure 3 per se, but rather learns to recognize the clustered patterns in the array of pixels in the image.[43] The input data in the neural net—and consider that in other instances, this could be anything: images, video, biometric templates, social media text—is assigned a series of probability weightings for its significance, with the output

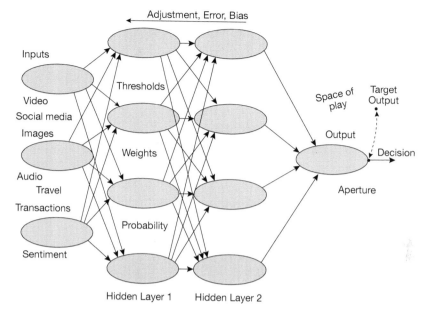

Figure 2.5   A representation of the arrangements of a deep neural network.

of each neuron to the next "hidden layer" dependent on whether the weighted sum of inputs is less than or greater than some threshold value. Each layer of the neural network detects some discrete aspect of the figure. As the computer scientists describe image recognition, "the learned features in the first layer" detect "the presence or absence of edges," with the second layer "spotting particular arrangements of edges," and the subsequent layers "assembling motifs into larger combinations that correspond to parts of familiar objects."[44] The recognition of edges, motifs, and familiar arrangements is not designed into rules by a human engineer but is definitively generated from the exposure to data. To be clear, this spatial arrangement of probabilistic propositions is one of the places where I locate the ethicopolitics that is always already present within the algorithm. The selection of training data; the detection of edges; the decisions on hidden layers; the assigning of probability weightings; and the setting of threshold values: these are the multiple moments when humans and algorithms generate a regime of recognition.

Adjusting the Features: The Variability of What Something Could Be

The computational problem of how to recognize people and things has become of such commercial and political significance that computer scientists enter their experimental algorithms in competitive image recognition contests. One particular algorithm, the AlexNet deep convolutional neural network, won an image recognition contest in 2012 and has become the basis for multiple subsequent commercial and governmental recognition algorithms, with the scientific paper cited more than twenty-four thousand times.[45] The AlexNet gives an account of itself—in the terms of a partial account I am advocating—that manifests just how its output is contingent on its exposure to data features, and the series of weightings, probabilities, and thresholds that make those features perceptible. As I recount something of how the AlexNet algorithm does this, I would like you to consider that if an algorithm is deciding "Is it a leopard?" or "How likely is it that this is a shipping container?" then it is also deployed to decide "Is this a face?" and "Is this face the same face we saw in the street protest last week?"[46] Understood in this way, the regime of recognition is political in terms of both arbitrating recognizability and outputting a desired target that is actionable.

As the computer scientists who designed AlexNet describe the relationship between recognition and cloud data, "objects in realistic settings exhibit considerable variability, so to learn to recognize them it is necessary to use much larger training sets."[47] The AlexNet CNN was trained on 15 million images, each image labeled by a human via Amazon Mechanical Turk's crowdsourcing labor tool.[48] Figure 2.6 shows eight test images for the algorithm, with the five labels considered most probable by the model assigned beneath each image. The algorithm is able to recognize a previously unseen image of a leopard or a motor scooter because the feature vectors of the image have close proximity to the gradients encountered in the training data. The capacity of the algorithm to recognize an incomplete creature at the edge of the frame (the mite) is considered to be a major advance in neural nets for image recognition. Where the algorithm failed to recognize an entity—the grille, the cherry—the scientists refer to the "genuine ambiguity" of which object is the focus of the image, and where the patterns of edges are occluded (e.g., the Dalmatian's spots and the cherries). To be clear, the logic of the AlexNet algorithm is that if one exposes it to sufficient data on the variability of what a leopard could be, then it will learn to anticipate all future instances of leopards. Indeed, deep neural nets are exposed to infinite variabilities—voting behaviors, faces in crowds, credit histories, kidney tumors, social media hashtags—to recognize the feature vec-

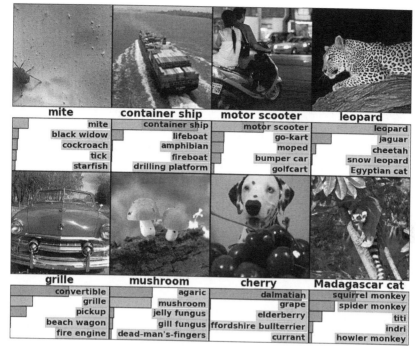

Figure 2.6 Test images of the AlexNet image recognition algorithm. Krizhevsky, Sutskever, and Hinton, "ImageNet Classification."

tors of all future instances. Whether someone or something can be recognized depends on what the algorithm has been exposed to in the world. Since the algorithm makes itself—adjusts thresholds and weights, for example—through its exposure to a world in data, it is becoming the contemporary condition of recognizability as such.

Like the cloud chambers of chapter 1, the propositional arrangements of the neural net are instruments of mattering, methods for making some things matter more than others. To seek to open or to make transparent the black box of this arrangement would be neither possible nor desirable, for the arrangement *is* an important site of politics, the spatiality of the calculus being politically significant in and of itself. This is significant for the ethicopolitical interventions one might wish to make because, for example, it could never be sufficient to demand that facial recognition algorithms that fail to recognize black faces be trained on a greater variability of images. For the algorithm also learns how to afford weight or value to one pixelated part of an image over others (the Dalmatian and not the cherry, the edges of this face and not that one). Indeed, as one computer scientist explained to me, a neural net like AlexNet, with six or eight hidden layers, is too complex even for the designer of the algorithm to explain the conditional probabilities that are learned. "I might adjust the weighting in that layer," he explains, "and I know that this will change the output, but I cannot say exactly how."[49] As with the design of AlexNet, the computer scientists work with the essentially experimental and unknowable nature of the algorithm. They perceive the fractional changes in the output of the model as they adjust the weightings, working with the emergent and unknowable properties of machine learning.

Bias Can Be a Powerful Ally

When deep neural network algorithms learn, then, they adjust themselves in relation to the features of their environment. To be clear, to learn, they have to weight some data elements of a feature space more than others—they have to have assumptions about how the world is ordered. Notwithstanding the widespread societal calls for algorithms to be rendered free of bias or to have their assumptions extracted, they categorically require bias and assumptions to function in the world. Indeed, even the textbooks used by the next generation of computer scientists address directly that "there can be no inference or prediction without assumptions," particularly the assumptions of "the probability assigned to the parameters."[50] Thus, when a team of European computer scientists discuss how they might move the threshold for their neural net algorithm to recognize the likelihood of a person of interest (a future person, yet to arrive) being a "returning foreign fighter" and not a "returning aid worker" from Syria, they mean that they will adjust the sensitivity of the algorithm to particular elements of weighted input data, such as increasing the probability weighting of particular past flight routes.[51] While some of this adjustment of the threshold is done by humans, today much of it is invested in the power of the algorithm to adjust itself in and through the emergent properties of the data, understood as a feature space. The "we both" of Turing's reflections seems to reassert itself here in the accounts of adjustment given by computer scientists Yann LeCun, Yoshua Bengio, and Geoffrey Hinton of Facebook AI, Google, NYU, and the University of Toronto: "We compute an objective function that measures the error (or distance) between the output scores and the desired pattern of scores. The machine then modifies its internal adjustable parameters to reduce this error. These adjustable parameters, often called weights, are real numbers that modify the input-output function of the machine. In a typical deep-learning system, there may be hundreds of millions of these adjustable weights."[52]

Like the abductive methods of intelligence gathering I discuss in chapter 1, this computational method observes the effect of the calculation—or the output signal—and theorizes back to the adjustment of parameters, like the mechanical knobs on a calculating machine. The distance between an agreed target output, or desired pattern, and the output scores is the *error* or the *bias*. Significantly, for the algorithm, *error is distance*; it is the playful and experimental space where something useful or "good enough" materializes. There is nothing normatively wrong about error in a machine learning algorithm, for it is a reduction of difficulty and difference into a mere matter of distance. Likewise, let me be clear, bias and weighting are not negative things for an algorithm. They are, on the contrary, essential elements of learning, so that, in computer science, "bias can be a powerful ally."[53] My point is that one could never satisfactorily address the ethicopolitics of algorithms by calling for a removal of human or machine bias and a reduction of error because the machine learning algorithm would cease to function at this limit point. Bias and error are intrinsic to the calculative arrangements—and therefore also to the ethicopolitics—of algorithms. At root, the algorithm can never be neutral or without bias or prejudice because it must have assumptions to extract from its environment, to adapt, and to learn. It is, ineradicably and perennially, a political being. To begin from here is to begin from the idea that all machine learning algorithms always already embody assumptions, errors, bias, and weights that are fully ethicopolitical. In the adjustment of parameters one can locate a shifting terrain of the relations of oneself to oneself and to others. The output of the algorithm is but a mere numeric probability, fragile and contingent, so that a tiny adjustment of the weights in the algorithm's layers will radically change the output signal, and with it the basis for decision and action.

Point Clouds and the Robot's Grasp

To extract something from the features in a data environment, to anticipate and to act, is a critical computational problem for deep machine learning algorithms in production line robotics, surgical robotics, drones, and IED (improvised explosive device) detection. Across these diverse domains, the capacity to recognize the three-dimensional form of an object and to decide on the optimal action is a challenge that animates computer science. Indeed, the failure to recognize multidimensional and mobile forms—such as those of human organs, vehicles, or facial features—has been a common feature of many high-profile mistakes and accidents by machine learners. In the fatal Tesla autonomous vehicle crash of 2016, for example, one way to articulate the error would be to say that the CNN algorithms failed to recognize the profile of a white van against a pale sky as the vehicle turned across the Tesla's path. The probabilistic answer to the question "Is this a vehicle?" was, fatally, "no." Discussions among computer scientists regarding the causes of such accidents are revealing in terms of a persistent determination to locate the source of the fatal flaw and to annex the algorithm from its milieu. For example, one group of IBM scientists urged caution "not to blame the algorithm for a failure of the sensors, the ambient lighting, or the human operator."[54] My point, though, is that what the sensor can sense, or the operator can decide, is only meaningful in the context of how the neural nets arbitrate what the objection could be, what it could mean. Similarly, in a Volkswagen factory in 2015, a robot failed to recognize the outline of a human coworker on the production line, mistaking him for a car door and crushing him to death. At the level of machine learning algorithms and their regimes of recognition, the da Vinci surgical robots' failures to recognize the boundary delimiting kidney tumor from human organ is not dissimilar to the biometric facial recognition systems that closed automated border controls at an airport when the setting sun changed the ambient lighting conditions. In all these instances, the algorithms have learned from the features of the environment they have been exposed to. Sometimes events and sensors in the environment will present them with a set of input features they have not encountered previously, and their assumptions and weightings may lead to a *spring of action* that misrecognizes the target.[55] Is this an error? Or is error merely a matter of distance?

The contemporary advent of cloud robotics has sought to address this problem of the limit point of exposure to features in a multidimensional environment. Cloud-based robotics, as we saw in the discussion of surgical robots, circulate data and aggregate computational power across a distributed system of machine and human learning. Just as the surgical robots are no longer limited to the data and computation stored within a bounded system, so the cloud-based intelligence system I discuss in chapter 1 recognizes its targets from exposure to data and analytics methods across borders and jurisdictions. Where machine learning intersects with cloud computing, the neural network algorithms are exposed to features from a vast archive of cloud data, including the Point Cloud Library of open source 2D and 3D images.[56] A point cloud is a set of topological data points mapping the 3D space of objects. Computer science research in cloud robotics is addressing the question of whether exposure to a vastly increased volume of point cloud data on objects can optimize the neural network's capacity to learn how to recognize and to act. Consider, for example, the computer science team at UC Berkeley's Automation Sciences Lab, whose research into cloud robotics is funded by the NSF, Google, and the US Department of Defense. Presenting their Dex-Net 1.0, or Dexterity Network algorithm, the Berkeley scientists experiment with CNNs to optimize the capacity of a robot to recognize and grasp a range of objects. The algorithm represents an advance on image recognition technologies such as AlexNet because it recognizes 3D objects from multiple viewpoints, and it outputs an optimal action based on this recognition.[57]

The Dex-Net algorithm is trained on an archive of Google point cloud data on "3D object models typically found in warehouses and homes, such as containers, tools, tableware, and toys."[58] The neural nets are learning to recognize the object's geometry and topology and then to optimize the robot's capacity to reach out and grasp the object. In a sense, the algorithm is asking a two-step question—What is this object? and How can it be most effectively grasped? This shift toward CNNs that can recognize *and* optimize an action is absolutely critical in the advance of robotics in manufacturing, medicine, and the military. The scientific papers on Dex-Net show something of the logics at work in coupling regimes of recognition to what I call a spring of action. When the Dex-Net algorithm is exposed to a training dataset of one thousand 3D point cloud objects (in figure 2.7, a household spray bottle), it is not able to find a "nearest neighbor" object that will allow it to recognize the query object. When the algorithm is exposed to the point cloud features of ten thousand objects drawn from the Point Cloud Library, however, it finds two proximal objects, or nearest neighbors, allowing it to recognize the object and optimize its grasp.

Though at first glance the Dex-Net's machine learning may appear as though the cloud is supplying "big data" volume to the algorithms, in fact the process of reduction and condensation I describe in chapter 1 is also taking place here. As the computer scientists propose, the significance of the "cloud-based network of object models" is actually to "reduce the number of samples required for robust grasps" and to "quickly converge to the optimal grasp."[59] Put simply, the Dex-Net algorithm is better able to condense and filter out the occlusions to recognize the most similar object. Each of the ten thousand cloud-based objects is prelabeled with 250 parallel-jaw robot grasps, each weighted with a probability of a successful grasp. "The goal of cloud robotics," as the computer scientists explain, is to "pre-compute a set of robot grasps for each object" so that "when the object is encountered, at least one grasp is achievable in the presence of clutter and occlusions."[60] Understood in these

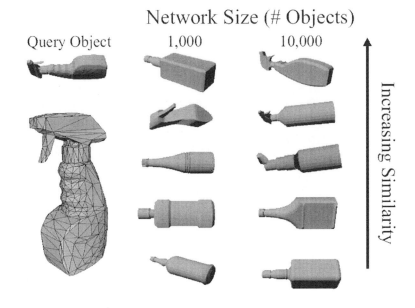

Figure 2.7 Point clouds for object recognition and grasp optimization.
Mahler et al., "Dex-Net 1.0."

terms, the weightings and probabilities of the point cloud make it possible to precompute something, so that the encounter with the unknown object can always yield an action that is optimal. With the point cloud, an algorithmic system does not only ask "Is this a face?," "Is this a bottle?," "Is this a military vehicle?," but rather it has already precomputed an optimal action in relation to the topologies it has encountered.

Precomputation: "The Fundamental Thing Is We Know What Good Looks Like"

On the horizon of research in neural net algorithms for robotics, then, one finds this notion of *precomputation* to make an action achievable amidst *clutter* and *occlusions*. Precomputation captures the neural net computational problem that extends from the recognition and grasp of the shape of a human organ amid "noisy" surgical data, to the recognition of a civilian body in the screened occluded data of the drone. One computer scientist, who now designs algorithms for human gait recognition, described to me how he had been "working with surgeons modeling the perfect operation," this human-algorithm collaboration itself deploying point clouds to precompute the trajectory of "a perfect procedure."[61] He has developed his algorithms for multiple applications, with each iteration another step in optimization: "The fundamental thing is we know what good looks like," he explains, "and then if you're doing anomaly spotting, you can see when something's wrong." For example, when the online gambling platform BetFair sought algorithms to recognize and act on patterns of addition, the designer suggests, "that's something we can spot because we know what addicted play looks like." This malleable normative assumption of "what good looks like" or "what addictive play looks like" is generated in and through the algorithm learning in a feature space.[62]

Precomputation implies that some sense of what is a "perfect procedure" or "at least one achievable grasp" action is always already present within the algorithm as such. The ethicopolitics of machine learning algorithms like AlexNet and Dex-Net is in the bias, weights, thresholds, and assumptions that make recognition precomputable. To precompute is to already be able to recognize the attributes of something in advance, to make all actions imaginable in advance, to anticipate every encounter with a new subject or object, a new tumor or terrorist, by virtue of its proximity to or distance from a nearest neighbor. The condition of possibility of the algorithm's action is its exposure to an archive of cloud data, condensed via the infinitely malleable value system of weights, probabilities, thresholds, and bias.

This is a pressing problem of the politics of algorithms in our contemporary moment. All our handwritten digits; all our online data traces; the biometric templates of our facial geometry; the point clouds of household objects and military hardware; all the movements of the hands, eyes, and bodies of surgeons, pilots, soldiers, consumers, production line workers: these are the teeming conditions of possibility of the machine learning algorithm. We are it, and it is us. We could never stand outside it, even if we might wish to. Each of the data fragments that enters the point cloud has a part to play in the learning. Whose is the grasp that caused the injury? Which of the 2.5 million objects in the archive became the nearest neighbor? Which of the possible 250 grasps for each object? Which of the many tens of thousands, or millions, of cloud-derived probabilities was responsible for the grasp that intuitively decided to pull the trigger, so to speak?

The harms inflicted through machine learning are not located primarily in the ceding of human control to machines, as is so often assumed in the ethical and moral debates on algorithmic decisions. Indeed, as we have seen via the surgeon who learns to reach and touch differently with her da Vinci robot, what it means to be human is significantly transformed in and through the machine learning algorithm. To appeal to the human as locus of ethics, then, is to appeal to a being already entangled with new forms of knowing and learning. The principal harm, in contrast, is manifested instead in a specific threat to a future politics. The tyranny of proliferating machine learning algorithms resides not in relinquishing human control but, more specifically, in reducing the multiplicity of potential futures to a single output. The claim to precompute the future, or to know "what good looks like" at the border, in the operating theater, in the economy, forecloses other potential futures. To be clear, the neural net does not reduce multiplicity as such. After all, as I have outlined, the spatial arrangement of the neural net algorithm contains within it multiple probabilities, infinitely adjustable weights whose emergent effects can never be entirely known, even to the designer. The finite elements of each hidden layer of the neural net, one might propose, contain within them infinite possible correlations to other elements. The spatial arrangement of the neural net does not foreclose alternative readings, different arrangements of what or who matters and what or who does not. Crucially, however, at the point of action, this intrinsic multiplicity is reduced to a single output. The insistence on a single output is the algorithm's orientation to action. Though I am reminded by the computer scientists that the output need only be between 0 and 1, and that there are infinite numbers between 0 and 1, there is nonetheless a single numeric output. Let us not forget that the algorithm's output signal lies behind the risk score at the border, the credit decision, the target assessment of the drone, and the decision on sentencing, detention, or the incipient dangers of a gathered protest on a city street. It is as though all the many potentials held in parallel, simultaneously distributed across the layers of the neural net, could never have been. With the output of the machine learning algorithm, one might say, things could never have been otherwise. The output is a probability whose value is transformed by the smallest of adjustments in the parameters of the model. And yet, nonetheless, all political uncertainty is rendered tractable on the horizon of the action triggered by this single output.

At this point, one might reasonably ask how giving such an account of the contingent politics of machine learning algorithms is of any possible critical use. How might a cloud ethics work with the incompleteness, the undecidability, and the contingency of the algorithm's space of play? If one wants to inquire whether a given algorithm is responsible for a flash crash in the financial markets, or if one seeks a human rights law adequate to the task of holding autonomous weapons or autonomous surgery to account, then some ethical grounds might be considered essential—or at least some method for accountability. As Michel Foucault proposes in his discussion of ethics, however, what may be necessary is not to appeal to grounds or to the juridical domain of statutes, but rather to "ask politics a whole set of questions that are not part of its statutory domain."[63] A cloud ethics must be capable of asking questions and making political claims that are not already recognized on the existing terrain of rights to privacy and freedoms of association and assembly. Cloud ethics belong properly not to the individual as bearer of rights, but to the many touch points and data fragments that are aggregated from the relations between subjects and objects. Thus, a cloud ethics must be capable of asking questions such as How did that Dex-Net algorithm weight the probability of that future grasp?; Why did the training data teach the algorithm to recognize this and not that object amid the occlusions?; How was the distance between target and output signal (bias) used as a space of experimentation?; and, In outputting that score, what were the traces of the rejected alternative weights and parameters? Such questions are necessary and urgent, even and perhaps essentially when they are unanswerable. The unanswerable questions reawaken the multiplicity that was, in fact, always present within the machine learning algorithm. All the many contingencies and alternative pathways are reopened, and the single output bears the fully ethicopolitical responsibility for the actions it initiates. The processes and arrangements of weights, values, bias, and thresholds in neural nets are, I think we can safely say, not part of our statutory political domain. And yet, I suggest that they must be presented as questions and political claims in the world.

Part 2

Attribution

The Uncertain Author
Writing and Attribution

I always think the notion of the fork in the road is very important when you are creating narrative, because you are continually coming to forks. You don't exactly know where you are going. . . . This is a state of uncertainty, or in terms of the modern physics, indeterminacy.
    —John Fowles and Susanna Onega, *Form and Meaning*

Source Code

In October 2017 the New York City Council proposed a new local bill—the Algorithmic Accountability Bill—conceived as a means to render the city's algorithms accountable for the significant effects their use has on the life chances of the people of New York City.[1] The city council's attempt to secure the accountability of algorithms was a response to the racialized harms precipitated by algorithmic decisions in the criminal justice system, where automated risk calculations further entrenched the notion that young African American men posed a high risk of reoffending and should serve custodial sentences.[2] As in many other cases when an algorithm has caused a harm or injustice, the reflex response of the authorities was to seek out the origin of the algorithm and locate the source of its harmfulness. In the first iteration of the Algorithmic Accountability Bill, the text called for public and private agencies "to publish the source code" of all algorithms being used for "automated decision systems," "policing individuals," or "targeted services" and "imposing penalties." The sponsor of the legislation, council member James Vacca, recorded his concerns that automated decisions made by algorithms were discriminating against New Yorkers based on the characteristics of race, age, religion, gender, sexual orientation, and citizenship status. Announcing that his ambition was to seek "transparency as well as accountability," Vacca located the actions and effects of the algorithm in what he saw as the determinate and discriminatory agency of the source code.

Though the Algorithmic Accountability Bill was subsequently amended to protect the commercial interests of algorithm vendors, before it was passed unanimously in December 2017, the appeal to a *source code* for the algorithm represents a broader impulse to locate the origins of the algorithm, or more precisely, to trace the source of the algorithm to an originary act of authorship. Put simply, the source of the algorithm becomes attached to the unified body of its code, anchored in the modern notion of the author who writes the code. The responsibility the algorithm can be said to have—for discrimination, for errors, for mistakes and injustices—is thus secured via the attribution of the authorship of code. This appeal to the authorship of the algorithm character-izes the many contemporary demands for accountable and transparent algo-rithms or, as I have described it, demands for an encoded ethics for algorithmic decision systems. An encoded ethics reasserts the human author of mathemat-ical code and locates the control of algorithms in the reasoning authority of the author. In some instances, as in the New York City Council's scrutiny of source code, the appeal to authorship is made through the body of work, via the locus of an identifiable source code. Here, code is assigned a unity of the body of work in the same way that authorship produces a unity from an oth-erwise errant and distributed text. In other manifestations, an encoded ethics is sought via an appeal to the writers of algorithms to adopt an ethical code of conduct that would secure the moral grounds for their writing of code.[3] Again, the location of ethics in the reasoning subject is mirrored by the irrevocable binding of the author to the body of work. Responsibility for the onward ac-tions and effects of the algorithm is imagined to be secured through the disclo-sure of authorship. Moreover, the call for accountable and ethical algorithms binds together not only source code and author but also data sources and au-thorship, so that it is said that "designers must make their data sources pub-lic."[4] A semblance of accountability for how the algorithm comes into being is found in the disclosure of the data sources from which the author writes. In sum, through the appeal to authorial sources and ethical design, much of the contemporary debate on algorithmic accountability is imbuing the algorithm with a specific form of what, with Michel Foucault, we might call an "author function."[5] This novel form of algorithmic author function has profound con-sequences for how one understands, and responds to, the ethicopolitical rela-tions forged through the writing of algorithms.

If the source code is the place where the logics of the algorithm can be found and disclosed to the world, then what exactly is the nature of this source? As I explain in chapter 2, in the context of how machine learning algorithms learn to recognize, the algorithm's ways of being in the world are not all present in the code. Indeed, the effects of the algorithm are generative and emergent so that the source code is never complete. Put differently, the source is continually edited and rewritten through the algorithm's engagement with the world. To understand the latent discriminations of the New York City algorithms necessarily involves engaging with its reincorporation of new entities into its code. In this sense, the sources and authors of the algorithm are continuously modified and elaborated through its actions in the world. To make its actions accountable via the source code is to misunderstand how what is meant by the category of *author* is reformulated in and through the algorithm. The authorship of the algorithm is multiple and distributed, extending across the attributes of one data population to act on the propensities of another. We are, each and every one of us, enrolled in the practice of the writing of algorithms. Through the practice of writing, we enter into new ethicopolitical relations with ourselves and with others. So, what is an author, to restate Foucault's famous question, in the context of algorithmic authorship?[26] In this chapter I explain the limits of locating an ethical response to algorithms in the authorship of source code. In place of the search for an origin of algorithmic harm, I propose that the actions of algorithms are producing new articulations of the author function that already possess politics and ethics. This already existing ethicopolitics of authoring algorithms could form the starting point to a rather different proposition than the search for accountability and transparency of sources. Understood in this way, the author function does not serve as the locus for accountability but is implicated in the authorization of algorithmic decisions. Though the desire to scrutinize source codes for intrinsic bias or discrimination seeks to control or regulate the decision-making authority of the algorithm, in fact the attribution of authorship is authorizing all kinds of renewed algorithmic decision.

I suggest an alternative formulation in which the ethicopolitics of algorithms dwell not in the originary responsibility of authorship but instead in the act of writing itself. Notwithstanding some of the distinctions that have been made between the writing of algorithms and the writing of literary text, I am mindful of the ways in which the notion of authorship operates at specific historical moments and across the boundaries of science and literature.[7] I draw on the reflections of novelists on the difficulties of generating an account of their authorship, not because there is a direct analogy between writing novels and writing algorithms, but because all forms of writing necessarily open onto an indeterminate relationship to oneself and to others. As Vicki Kirby proposes in her feminist reformulation of deconstruction, there "is no 'every-thing' that pre-exists the relationality that is the scene of writing."[8] Taking seriously Kirby's argument that the elements of the scene are already present in the instantiation of writing, and that writing enrolls human and nonhuman authors, the writing of algorithms also involves the instantiation of a scene. If writing is also world making, then to give an uncertain, partial account of oneself as author is to open onto the difficulties of the *I* who makes decisions and chooses pathways as well as the other readers and writers who will follow different forks and turns in the future. As the novelist John Fowles and his co-author Susanna Onega reflect in the opening to this chapter, the practice of writing involves encountering "forks in the road" at which it is never certain what might take place as a result of the decision at the branch. To write, as Fowles experiences it, is necessarily to encounter an uncertain and indeterminate future. Though the writing of algorithms and the writing of literature is not at all the same thing, of course, the bifurcated "forks" in the algorithm's calculative architecture do involve a profound uncertainty of authorship and its future effects. Conceiving of writing as a practice in which scattered elements and contingent forks in the road are gathered and unified, the writing of algorithms does constitute something akin to an ethicopolitical writing practice that involves "manifesting oneself to oneself and to others."[9]

Returning to my account of the ethics of algorithms located in the relations of self to self and self to others, writing algorithms shares with the novelists' accounts the impossibility of a determinate and grounded authorship. In place of locating ethics in the individual subject of the author and the authorial source of the work, the ethics of algorithm emerges from the composite and collaborative practices of writing in which we, and our scattered data elements, are also implicated and engaged. To propose that algorithms dwell within us, and we within them, is also to suggest that through these relations an ethics of writing algorithms is forged. In this reformulated approach to authorship and the algorithm, it would be insufficient to respond to the violences and injustices of an algorithm by "scrutinizing the source code" because the act of writing these algorithms substantially exceeds that source, extending into and through the scattered elements of us and the multiple data points that gather around us.

The Author and the Attribute

The computer science field of natural language processing (NLP) has become of intense interest to those developing experimental deep learning algorithms. Beginning in the mid-twentieth century with models for automatic language translation, NLP for much of its history has used basic decision tree algorithms and rules-based techniques to render human language computable.[10] In my book *The Politics of Possibility*, I traced Rakesh Agrawal's rule-based method of text mining as it developed from the mining of customer retail basket data in the early 1990s to the mining of data for counter-terrorism after 2001.[11] With the availability of volumes of cloud-based unstructured data in the twenty-first century, however, NLP methods have shifted from the rules-based deriving of semantics from text to deep learning algorithms generating meaning from their exposure to a corpus of data derived from speech or text.[12] The significance of this transformation from rules-based determinants to generative models of the meaning of a text is that the algorithms are no longer seeking exact one-to-one matches of semantic meaning; rather, they are building predictive models of what might come next. They are, in short, inferring future worlds from their exposure to text. Thus, for example, corporations are increasingly shaping their customer relations management around models of customer propensities built from the NLP of social media text. Similarly, as I detail in chapter 1, security and intelligence agencies are using NLP of Twitter and Facebook data to build predictive models of future threats. These computational problems are not understood primarily in terms of the recognition of the semantic meaning of text, but instead are concerned with computing human context, meaning, patterns of behavior, and possible futures.

This manifest desire to find computational methods for futures-oriented inference from text has witnessed a curious new relationship emerging between science and literature (and between data and narrative forms), where literature becomes a corpus of contextually nuanced data from which a deep learning algorithm is trained to recognize the meaning of human language. An example of how computer science is turning to literature to teach machines how to build predictive models of meaning is led by the computer scientist and author of the notion of a coming singularity, Ray Kurzweil. Google's Natural Language Understanding Research Group (note *understanding*, not *processing*) is teaching algorithms how to infer the subtleties and nuances of human language.[13] The Google team trained a deep neural network algorithm on a corpus of data comprising the literary works of one thousand deceased authors, from William Shakespeare to Daniel Defoe, and from Virginia Woolf to Herman Melville.[14] The neural net algorithm was reported by the scientists to have discovered the literary style of particular authors from the patterns within their body of work. In fact, the algorithm had done what most machine learning algorithms do: it had clustered the text according to the features in the training data and defined these clusters in terms of the *attributes* of the author's body of work. A similar method was used in 2013 to attribute the pseudonymous novel *A Cuckoo's Calling* to the author J. K. Rowling, with the attributes of nom de plume Robert Galbraith's writing correlated to Rowling's.[15]

Once recognized and learned, the attributed features of an author's text become a computational means to identify the future attributes of as yet unknown texts. This attributive capacity allows algorithms to anticipate future features of a text, whether a person's financial patterns or their writing on social media. During its development and testing, the Google NLP algorithm was given an input sentence from a novel not present in the training dataset, with the target output defined as a probability that this sentence was a match to a given second sentence. At this point the data was unlabeled—the algorithm was not given the name of the author—yet it made a correct prediction in 83 percent of the input queries. As the Google team account for their experiments, their guiding question was, "Can we build a system that can, given a sentence from a book and knowledge of the author's style and 'personality,' predict what the author is most likely to write next?" In a later phase of the experiment, the team built a generative model that was able to respond to questions in the style of a specific author, said to be predicting the future text of "dead authors" from "beyond the grave."[16] In short, the algorithmic technology of the attribute is not about identifying the qualities of an entity in the present, but about rendering something attributable into the future.

The significant point here is that the author function in contemporary deep learning algorithms is serving a futures-oriented capacity to associate entities through their attributes. The name and source "identity" of the author may be unknown or anonymous, but nonetheless the algorithm is deriving an author function from the clustering of attributes. Put simply, what matters to these deep learning algorithms is not primarily the naming of an author but the generating of a set of attributes that could be latent in any other entity. As I have similarly described the border-control algorithms that can act on an individual indifferent to their naming or identification—as a set of associations—so natural language algorithms need only to be able to associate the attributes sufficiently to say "this sentence X matches the next possible line Y with probability P."[17] Even where the strict naming of the author is not possible, the algorithm can work productively with the attributes of the body of work.

One might reasonably ask what political or ethical significance could possibly be attached to this distinctive learning of attributes via clustering techniques. Yet, the computational capacity of NLP algorithms lies behind many of the apparently autonomous decision-making systems that are of ethical and political concern. From the extraction of text from images of public protests, to the analysis of the sentiment of social media text for detecting the propensities of a person, these algorithms are concerned precisely with the inference of possible futures that could be associated with a set of attributes. The Google team explains the motivation for the one thousand authors project as "an early step towards better understanding intent."[18] Given the attributes of a given cluster within a corpus of data, what is the future intent? The predictive power of the attribute is located in the capacity to infer meanings in the future. Thus, for example, how does the intent of one line of text—"I will go and sort him out" (to seek revenge and to threaten violence)—differentiate itself from the intent of another that is semantically identical—"I will go and sort him out" (to offer support and to care for someone's troubles)? It is exactly this kind of subtle distinction between ostensibly identical text that is sought by the algorithm's designers. For the algorithm, the capacity to discern difference is derived from the extracted features of written style and genre as a set of relations. Contemporary algorithms to be used across domains, from credit card fraud to counter-terrorism, are being trained to understand future intent through the attributes of style and genre.

As the Google experiments with authorship and inferring future intent suggest, though the algorithm does not require a definitive named author per se, it does seek to establish an author function in which the attributes of a corpus of text yield the writing of the next line into the future. As Michel Foucault proposes, "Criticism and philosophy took note of the disappearance—or death—of the author some time ago."[19] Understood in this way, the disappearance of the author is not a recent phenomenon but is intrinsic to the act of writing, in which "the writing subject continually disappears."[20] And yet, at the very moment when it seems the author has disappeared, just when it might seem tempting to appeal to the author by name and demand responsibility for what is written (as New York City Council members did in their call to scrutinize the source code), "a certain number of notions that are intended to replace the privileged position of the author actually seem to preserve that privilege."[21]

Here, in Foucault's discussion of how the author function continues to live and thrive amid the apparent death of the author, I find some resources for understanding what is happening when algorithms are learning from a corpus of text to infer future propensities. For Foucault, even when the author appears to have lost status, the body of work stands in for authorship so that "the word *work* and the unity that it designates are probably as problematic as the status of the author's individuality."[22] Thus, even when a named individual author is absent (almost always the case with algorithms), the notion of a *work* signals a stylistic unity of characteristics of particular authorship. For this reason I understand the *source code* of an algorithm to invoke the author function of designating a unitary whole even when this apparent whole is composed of heterogeneous fragments of data and past writings. Similarly, the corpus of literary data from dead authors, used by Google to train algorithms, is afforded the unity of a "work": a unified collection of styles, patterns, and behaviors from which future attributes can be derived. What categorically does not count as part of the authorial work in the Google NLP, however, is the actual work of the algorithm as it iterates back and forth, adjusting probability weights and thresholds and modifying its own code. This annexing of the writing of source code from the back-and-forth iterations of an algorithm rewriting has profound implications for ethics. The designers of facial recognition algorithms for policing, for example, commonly testify that bias and discrimination reside in the human operator and not within their code. When one follows the iterative writing of the algorithms in practice, however, one sees that the actions of a human operator are in close communion with the algorithms, so that each decision also recalibrates the model. In short, to annex the writing of code from the iterative writing in the world is to seriously miss the ethicopolitical life of algorithms. As Foucault suggests, the preserved dominance of the author function hinders us from "taking the full measure" of the author's disappearance and attending to what is happening in the gaps and spaces of that disappearance.[23]

Foucault's writings on the subject of authorship also afford us a glimpse of what might be happening when texts are grouped together for machine learning. In the Google experiments, what matters for the algorithm is not primarily who the author is, but rather what attributes make a set of texts part of a cluster defined by degrees of similarity. For Foucault, the author function serves a kind of "classificatory function" in that it "permits one to group together a certain number of texts, define them, differentiate them from and contrast them to others."[24] Though the form of algorithmic author function I wish to depict is not classificatory in the strict sense of grouping things by similarity, it does serve a clustering function in terms of finding structure in data and learning what normal or abnormal looks like. In this sense the author function retains its privilege to find a resolution of differences and tensions in the "authentication of some texts by the use of others."[25] To be clear, it is profoundly misleading to appeal to the author of the algorithm or to the source code as the locus of responsibility because the algorithm's power lies in its capacity to promise to resolve incalculable differences via the unity of its outputs. To invoke the call for an identifiable and responsible author is not only insufficient as critique, it also risks amplifying the already present ability of the algorithm to "permit grouping together" so that "incompatible elements are tied together," with "all differences having been resolved."[26] In place of the ethicopolitical call for securing responsibility via authorship, then, I am interested in deepening our sense of how the algorithm ties together incompatible elements to appear to resolve the politics of difference.

Authorship and the Algorithm

In Foucault's discussion of the author function in sciences and literature, he notes the shifting historical modes of authorship in the eighteenth and nineteenth centuries. Remarking that in European history, "it has not always been the same types of texts that have required an author function," he reminds us that there was a time when literary narratives were widely circulated and accepted without an author because of their ancient and received origins. Just as literary texts could be authenticated anonymously, so "those texts we would now call scientific" were accepted as truth "only when marked with the name of their author."[27] A transformative reversal of the author function took place in the eighteenth century so that "the author function faded away in science," where truths were founded on proof, theorem, axiom, and law. Simultaneously, literary texts came to be accepted only when afforded a particular form of author function: "We now ask of each poetic or fictional text: From where does it come, who wrote it, when, under what circumstances, or beginning with what design? The meaning ascribed to it and the status or value accorded it depend on the manner in which we answer these questions. And if a text should be discovered in a state of anonymity—the game becomes one of *rediscovering the author*."[28]

The significant point is that the notion of authorship does not belong naturally to the literary world in a way that it does not belong to science.[29] Nor is it inappropriate to use the concept of authorship in relation to the writing of algorithms, for the category of the author has functioned differently across domains of knowledge and at different historical moments. In our contemporary moment, the author function is being rearticulated in a novel way in relation to algorithms. As computer science has begun to embrace the emergent and intuitive faculties more commonly historically associated with literature, so the logic that science has shed authorship while literature has retained it begins to erode. Contemporary computer science is invoking just the kind of questions that Foucault ascribes to each poetic or fictional text. Indeed, the questions used to train deep learning algorithms follow closely those associated with literary authorship: From where did it come? Beginning with what design? What is the meaning ascribed? What is the value? The twenty-first century marks a moment when the specific form of the author function is once again changing its shape. "It is not enough to repeat the empty affirmation that the author has disappeared," writes Foucault, but "instead we must locate the space left empty by the author's disappearance, follow the distribution of gaps and breaches, and watch for the openings this disappearance uncovers."[30]

It is this distribution of breaches and openings that I wish to revalue in a reformulated author function of the algorithm. Indeed, Foucault signals what some of the breaches might be in the author function within mathematics. Acknowledging the plurality of self involved in the literary novelist's use of *I* in the first-person narrator's account, he points to the "scission" between the author and the "real writer," the "plurality of self" that opens up in all cases "endowed with the author function." Crucially, this plurality of self is specifically present also in "a treatise on mathematics," where authorship splits, divides, and disperses so that it "can give rise to several selves, to several subjects."[31] A significant opening, then, begins from the way the first-person *I* as a locus of authorial responsibility is actually always divided and dispersed. This is not only a matter of the dividuated subject, divided within herself, as is often documented in the accounts of digital subjects. The plurality of authorship also involves a dispersal across different human and nonhuman life forms.[32] For example, as the neural net algorithms animating robots become more sensitive to context and the attributes of behavior, they are modifying their own actions by learning from their interactions with other robots.[33] The algorithms are not simply the "authors of their own code" in the sense of locating authorship in an autonomous machine, but instead they are engaged in acts of writing and rewriting that unify otherwise scattered and heterogeneous elements.

So, when a deep neural net predicts a deceased author's next sentence, or when a New York City policing algorithm decides the attributes of a person at high risk of reoffending, within the author function teems a multitude. In attempting to find mechanisms of accountability for algorithms, our societies are searching for an authorial source of the writing of code. Yet, this desire for a first-person author as the locus of agency and responsibility deepens the power of the algorithm to present itself as a unified entity.[34] Simultaneously, it underplays the contingency of the algorithm and conceals the breaches that might otherwise appear in the constraining figure of the author. In the following pages I propose a different approach to authorship and the algorithm, foregrounding the multiple and dispersed acts of writing that are its condition of being.

Writing and the Algorithm

To this point my argument has emphasized how the origins of algorithmic actions are commonly understood to reside in the unity of the source code. Anchored in the modern notion of the author, responsibility for the decisions, harms, and injustices of algorithms tend to be traced to the notion of a source code where errors or bias could reside. I have explained why this appeal to a source code sustains the status of identifiable authorship as the locus of responsibility, binding the author to the work and unifying the works as though homogenous and undifferentiated. At the very moment algorithms usher in novel forms of the author function, the resort to ethical frameworks for coding sustains the vision of a corrective "fix" in which society asserts control over the actions of the algorithm. In what follows I outline how understanding the ethics of algorithm requires a break with notions of authorship, and I develop instead concepts of writing algorithms that substantially exceed the method of tracing sources.

A place to begin to explore the distinction between authorship and writing is where the source codes of an algorithm were the subject of legal contestation but, once released, were followed by unexpected events that exceeded the scrutiny of the code. In the 2017 federal gun possession case of *United States v. Kevin Johnson*, presiding judge Valerie Caproni ruled that the source code of the New York Chief Medical Examiner's Office (CME) DNA analysis algorithm had to be revealed to the defense team.[35] The Forensic Statistical Tool (FST) used by the New York CME, in common with other commercial algorithms used to analyze DNA material at crime scenes, is an algorithm designed to make it possible to calculate the likelihood of an individual person's DNA being present at a scene, even when the DNA material is an incomplete profile and is mixed with the DNA of others. In effect, with these machine learning algorithms, incompleteness and uncertainty are rendered calculable as probabilities. The important difference between algorithms such as FST, TrueAllele, and STRmix—which have been used in many tens of thousands of criminal convictions in the United States and elsewhere—and traditional statistical methods for analyzing DNA material is precisely how the probability is generated by the algorithm.[36] The FST algorithm generates a *likelihood ratio*, which is a numeric probability that a *known suspect* individual contributed DNA to a mixed and incomplete sample as compared to the probability that an *unknown other subject* in the general population contributed that DNA. When there is an incomplete DNA profile as part of a mixture of DNA material at a crime scene, the algorithm compares that incomplete fragment with the probability that it was derived from the general population. As I explain in chapter 2 in relation to machine learning, this process means that the attributes of one population are used as the basis for calculating the probabilistic likelihood of the actions of another person.

In the case of the FST algorithm, as described in the forensic biology protocols of the New York CME, "the likelihood ratio value provides a measurement of the strength of support for one scenario over another, i.e., one scenario being that a known person contributed to a mixture versus the scenario that an unknown, unrelated person contributed instead."[37] In short, where the DNA profile of an individual cannot be identified from the materials at the scene—where the trace is incomplete or indeterminate—the likelihood ratio algorithm allows for partial samples to be "positively associated" with an individual by comparison of their attributes to those of a group of unknown others. A DNA *profile match* becomes, via the algorithm, a correlative and scenario-based DNA *positive association*.

There can be little doubt of the profoundly troubling implications of using such algorithms to convict individuals of crimes.[38] When one looks closely at what happened when the source code of FST was released to the defense lawyers, however, it becomes clear that the source of harm resides not in the code itself but in the iterative and heterogeneous writing of a likelihood ratio. Once one relinquishes the notion that the authorship of the algorithm resides in the code, following instead the writing of the algorithm, the code is not the authorial source of the algorithm's actions but merely one element in an iterative process of writing. Among the defense team expert statements—each of them concluding that the algorithm could have resulted in innocent people serving a prison sentence for serious offenses—one computer scientist, Nathaniel Adams, reported what he found when he examined the FST algorithm. His findings do not pertain to an error or flaw in the source code or even to a bias contained therein but specifically to the way in which the algorithm learns to write a likelihood ratio from a body of data. In a method that is resonant with the Google machine learning algorithm's prediction of the next sentences of deceased authors based on a corpus of literary data, the FST learns from "the databases accompanying the source code."[39] These databases include the training and validation sets from which the likelihood ratio model is generated. The algorithm modifies its output according to adjustments in the input data. Adams experimented with adjustments to the weights of the algorithm and demonstrated the contingency of the writing of a likelihood ratio when it is expressed as an output score. Of particular significance is that the algorithm's sense of what the unknown population looks like in terms of DNA is clustered along racialized attributes. Adams discusses the four "subpopulation" datasets used for the training of the algorithm—Caucasian, Hispanic, African American, and Asian—and he emphasizes the contingent effects of the "removal of a locus" from these subpopulations on how the algorithm learns to generate a likelihood ratio. In sum, the root of harm is not the authorial source code but predominantly the way the algorithm is written and how it writes itself—continuously editing, adjusting, removing, and iterating in relation to a corpus of data. The attributes of the DNA of a "general population" and an "unknown unrelated person" do not preexist the text but are written into being. As the computer scientist testifies in his expert statement, "Due to the complexity of probabilistic genotyping algorithms, seemingly small modifications can have unintended consequences and compounding effects that alter the behavior of the algorithms."[40] Let us not forget that these tiny modifications will change the output score of the algorithm and, therefore, change the likelihood of someone being present at a crime scene.

Though ostensibly the source code of the genotyping algorithm was of juridical and political concern, once the code was released onto the software development platform Github so that it forged new relations of possibility, what came to matter was not an imagined anomalous source code but the capacity of the algorithm to continue to write its relations to normalities, abnormalities, likelihoods, and probabilities into the future. Thus, the algorithm's relationships with data is one of writing the world, bringing into being a singular output—a likelihood ratio—from the disparate, heterogeneous, partial, and fragmented scattering of data elements.

In proposing that one could follow the breaches and openings of the author function, then, I am making the iterations, contingencies, and likelihoods of the algorithm the focal point of ethical concern. The calls for access to the source code of algorithms are a means to try to read how a likelihood ratio is written. Yet, scrutinizing the code itself would reveal nothing of the racialized, prejudicial outputs of the algorithm that are written via training data, validation data, and the feedback loops of experiments in the field. Indeed, the Github platform is itself a form of distributed and iterative writing in which multiple developers contribute to the rewriting and editing of software. If one is to take seriously the practice of writing as an ethical resource in responding to algorithmic injustice, then how does one follow this practice of writing? What kinds of strategies can allow us to pursue the writing of the algorithm in place of its authorship? What might the effects of this foregrounding of writing over authorship be? Inspired by the reflections of novelists on the practice of writing, in what follows I suggest a series of openings onto the writing of algorithms: bifurcated paths; *I* as a character; and the obligations of the reader.

Bifurcated Paths

In an essay on the practices of science and literature, John Fowles reflects on his experience of what happens when he writes. "An important term—at any rate for me, in my own practice," writes Fowles, "is the fork (as in a path), by which I mean a fairly continuous awareness of alternatives, both learned (remembered) and fortuitous (wild), in what is done."[41] In Fowles's terms, the practice of writing involves the encountering of contingent moments, or forks in the path, where a decision is taken amid the continuous awareness that alternatives exist. To take the alternative path at the fork, Fowles seems to suggest, will change the narrative of the novel, modify its future course and potential ending. Indeed, for the novelist, this partiality and opacity concerning what is to come in the future characterizes the very value of writing. Reflecting on the writing of William Golding, whom he calls a "master fabulator," Fowles notes a sustained "lack of expectability" that does not conform to an expectation of style but works against the grain of the literary rhythms and patterns of the time.[42] There is an almost ethical emphasis in Fowles's notion of writing against expectation and with uncertainty, an ethical emphasis that extends from literature to the sciences. "Any true scientific knowing," writes Fowles, "is always, like feeling, only partial."[43] The fork in the path of science, as in literature, is taken with partial knowledge and in the face of profound uncertainty: "Cruel, painful, even death-dealing though it may be, we could not live in this world without its sheer incomprehensibility."[44]

Inspired by Fowles's reflections on his practice of writing, an important distinction to be made is that between a claim to authorship that precisely seeks to secure against uncertainty and produce the expected (where even the next lines of a deceased author can be expectable and inferred from their style), and a form of writing that opens onto uncertainty and the unexpected. The writer as fabulist, as Fowles describes Golding, and as Gilles Deleuze proposes, is "someone who creates their own impossibilities, and thereby creates possibilities."[45] In this way the practice of writing is also a practice of fabulation, in which the idea of the author is continually undone by the writing, by the impossibility of knowing in advance where the forks in the path might lead.

This encountering of forks in the path is exactly what characterizes the writing of algorithms. Once one has relinquished the idea of authorship and focused on the writing, all algorithmic arrangements involve forking paths, from the branches and leaves of decision trees to the differential paths through the neural network. Where for the writer of the novel the fork in the path may represent a decision amid alternatives in the context of uncertainty, so for the writing of algorithms the fork is a weighing or adjustment of probabilities and likelihood. What could be made to matter, in terms of the ethicopolitical response to algorithms in society, is to trace these branching points: not as a means of securing the knowable source, but precisely to confront the profound contingency of the path. If a forensics algorithm outputs a 0.62 probability of a young African American woman's DNA being present as a fragment in a mixture at a crime scene, then how does the adjustment of the weights at each fork change the dissolution of alternative paths, other decisions not taken? Which other unknown populations had to be enrolled for this likelihood to be followed while another was discarded? How can the writing of the algorithm be reopened at the points where there could be renewed awareness of what cannot be attributed? What kind of interventions could be made at the moment of condensation, when something brimming with other potentials is reduced to one?

"I Is Simply Another Character"

John Fowles, writing in 1967 on the difficulties of writing his then-unfinished novel *The French Lieutenant's Woman*, recounts how the first-person narrative does not consist of "the 'I' who breaks into the illusion" of fiction, but rather engages the "I who is part of it."[46] The *I* is not the authorial voice who enunciates and controls the narrative, then, but is simply a character among the many who together write the narrative. The *I* who writes the first-person commentaries in Fowles's unfinished novel is not "my real 'I' in 1967" but instead "simply another character" in the iterative play of characters in the novel. Even the first-person account, then, does not secure the authoring subject and take them outside the fabulation of the writing of the novel. On the contrary, as literature and philosophy have long proposed, writing is the opening of a space into which the author disappears.[47]

To conceive of the writer as simply another character in the novel is to interrupt the author function and its place in the sovereign economies of the ownership of text and the responsibility for the originary source of a text.[48] The novelists and philosophers of the late 1960s shared an interest in the capacities of writing to interrupt the authorial voice and open onto the writing of indeterminate and uncertain future worlds. It is no coincidence that in 1967, when the novelist John Fowles displaces the author in his account of writing, the philosopher Jacques Derrida publishes *Of Grammatology* and defines the methods of deconstruction. In these traditions, writing always breaches the authorship of a text because it necessarily exceeds its context, iterating beyond the moment of being inscribed and enduring "even after the death of its author."[49] Reading the insights of deconstruction into my treatment of algorithms as arrangements, I find a profound contrast between an author function that witnesses algorithms predicting the next lines of a deceased author based on clustered attributes, and a writing that iteratively exceeds the *I* of the author. To bring to the fore the writing of algorithms is to unsettle the notion that one could locate responsibility for the harms of an algorithm in its source code. The algorithm iterates beyond the moment of its inscription, distributing the writing through multiple characters, from the training data to the back propagation of errors. This matters because any meaningful response to the actions of algorithms, as we saw in relation to genotyping neural networks, would need to relinquish the search for the source code and focus instead on the iterative writing back and forth between DNA data on the population and the making of a likelihood ratio. What would be sought in such a strategy would be the awareness of alternative paths, as Fowles suggests. There are multiple learned and "in the wild" features at work in the writing of algorithms, from the many fragments of past data lodged in the criminal justice databases to the feedback loops that adjust and modify the algorithm into the future.

"To Oblige the Reader to Help Form the Text"

In an interview in 1995, John Fowles reflects on the place of silences and omissions in literature. "One of the greatest arts of the novel is omission," he suggests. "I'm a deep believer in silence—the 'positive' role of the negative."[50] In the gaps and omissions of the text, as Fowles concurs with Samuel Beckett and Herman Melville, is the space where the reader enters "to help form and to experience the text" and where reading becomes a "heuristic process."[51] The relation between writer and reader, then, is a profoundly ethical one in which the writer must address an absent subject who will nonetheless modify the writing, and the reader cannot simply read according to the signs or rules of the text. As Simon Critchley poses the question, "How is the ethical relation to the other person to be inscribed in a book without betraying it immeasurably?"[52] In response, Critchley proposes that "ethics is first and foremost a respect for the concrete particularity of the other person in his or her singularity," so that "ethics begins as a relation with a singular, other person who calls me into question."[53] The writing of a novel, then, has what Fowles describes as an "ethical function," not because it lays down for the reader the principles or rules of how it ought to be read, but on the contrary, because it contains gaps and openings for the concrete singularity of the reader. The writer and the reader are necessarily and intractably tied together in the difficulty and obligation of forming the text. And there can be no critical outside to their relation, but always an immersed involvement in the processes of writing and reading. "One cannot stand on the bank," says Fowles. "One is willy-nilly in the stream. And by 'one' I mean both writers and readers."[54]

To be in the stream, then, is to be already involved in the ethicopolitical acts of writing and reading. When one hears of the automated analysis of *data streams*, it is exactly this imagination of already being *in the stream* that I consider to be so crucial. We do not stand on the bank and respond to the stream, but we are in the stream, immersed in the difficulties and obligations of forming the text. The broader implications of being in the stream as the writers and readers of algorithms are significant. One so often hears the claim that algorithms are illegible and unreadable, that they operate at scales and speeds in excess of human reading. But my point is that we should remember that algorithms are not unique, or somehow "outside the text." All acts of writing and reading, whatever the text, necessarily confront illegibility and the impossibility of reading.[55] We do not need new resources or technologies to prize open the unreadable algorithm. Instead we could begin from that unreadability as the condition of all engagements with text. It is something familiar to us all as social and political theorists. We are, in fact, well equipped to confront the difficulty of what seems to us an opaque text.

An ethics of algorithm could usefully respond to the specific ways in which algorithms address the others with whom they read and write the future. Critchley places his emphasis on the ethical address to the singularity of the other person, but the writing of the algorithm categorically recognizes no singularity or particularity. The algorithm learns to recognize only a set of differences from, or similarities to, the attributes drawn from another data population. The injunction for the reader to complete the text, then, can only find a response in the exposure to more data to write the text. In place of a search for the authorship of the algorithm, I am asking for a response to the concrete particularity of the other person who is obliged to help form the text. What this means in practice is that no person would be addressed as the "probable association" derived from a likelihood ratio algorithm. The algorithm writes the world into being in ways that oblige others to form the text, but as it does so, it necessarily also encounters a singular reader whose particularities enter the omissions in the text in ways that are unexpected.

Fabulation and the Algorithm

In the fifth of Hilary Mantel's series of BBC Reith Lectures in 2017, the novelist expresses the relationship between telling history and narrating a story. "We need fiction to show us that the unknown and unknowable is real and exerts its force," she argues, dispelling the notion that the facts of history are distinct from the fictions of her narratives of Tudor England in the novels *Wolf Hall* and *Bring up the Bodies*.[56] For Mantel, the writing of fiction confronts the reality of the profound unknowability of our world. The unknown nature of our actions, relationships, and intentions is the truth contained within the novel. As I have discussed through the essays of John Fowles, writing is distinct from authorship precisely because it must confront the force of the unknown, and not least the unknowable *I* that is the writing subject. I have sought not to diametrically oppose the writing of the novel and the writing of algorithms, but rather to distinguish the author from the practice of writing so that the sovereignty of the source code is opened up to the writing, of which it is one element. This writing—whether literature or code—necessarily and irredeemably confronts the force of the unknown and the unknowable, as Mantel describes it. What is at stake here is a rather fundamentally different response to the problem of the algorithm's accountability. Where the demand for ethical frameworks for code seeks to make the algorithm knowable and governable, my call for attention to the writing of algorithms demands a different mode of account.

To call for the algorithm to give an account of itself is not to demand disclosure or transparency, but precisely to express something of the unknowability of the algorithm and the future that it brings into being. In this sense, like novels, algorithms can be understood as fabulations, where the production of an output involves a process of working on a corpus or body of material.[57] "The fabulatory function," as Gilles Deleuze proposes, "strips us of the power to say 'I.'" To write is not to anchor a narrative in a determinate first-person source, but precisely to work with the incomplete and indeterminate traits of characters and their connections. Writing is fabulation because it projects into the future with "a life of its own" beyond the author's schema, a life that is "stitched together" and "continually growing along the way."[58] Of course, one could object that the algorithm actually produces completeness and a determinate future via a single output, an automated decision that goes out into the world of criminal justice, health risk, security, and war. Yet, this is precisely the paradox of the process of writing, of fabulating a text. The writing of the algorithm fixes a unity from scattered data elements, at that same moment fabulating new connections and traits, forging attributes that will attach to other beings into the future.

The task as I see it, then, is to find ways to amplify the fabulation of the algorithm, to enter the breaches in the writing so that the force of the unknown can be felt. I understand the writing of the algorithm to substantially exceed the writing of source code and to extend into the iterative writing, editing, and rewriting of composites of data, humans, and other algorithms. Writing in this mode is a "practice of the disparate," consisting of gathering "heterogeneous elements" so that traits, sensibilities, and attributes emerge from the composite.[59] To be clear, the practice of writing algorithms is located not in the individual author, but in the multiple and dispersed writers of the algorithm—from the data that quietly enters the training set to the person who experiences the full force of "high risk" attributes. For even the target of the algorithmic decision becomes a new set of data to be worked into the refining of weights and probabilities. If fabulation involves what Deleuze calls "strictly speaking a series of falsifications" generated by working on a material, then the algorithm appears anew.[60] Stripped of its authorial power to speak the truth and to produce objective outputs, the writing of the algorithm necessarily involves a series of falsifications. The algorithm is not false in the sense of untrue—it is indifferent to truth except in relation to data-grounded truths—but false in the sense of wrought from a set of materials that could be infinitely reworked.

To be clear, it is not my argument that the writing of algorithms is analogous to the writing of literature. On the contrary, I argue that the philosophies and social theories of writing contain a rich resource for considering all forms of writing in an extended sense, as a "network of traces."[61] All languages and codes, then, signify through differential traces, where a trace makes present something not present in the past. In common with all languages and codes, the algorithm writes to make present something that is not merely in the past data. Interrupting the notion that an algorithm's source code works on the material of what is in past data, the writing of algorithms fabulates the traits and attributes of a cluster so that it makes present what did not exist in the past. What is at stake here is that public discussion of the accountability of algorithms is concerned primarily with the problem that the source code of algorithms cannot be read. Yet, the unreadable and uncertain nature of all forms of writing has provided a core of our philosophical debates for many decades. Does the writing of algorithms really pose a problem for which we do not already have a rich vein of ethicopolitical resource?

The methods of deconstruction, of course, have never been confined to the writing of "text" in the form of a book, a script, or another body of written work. If, as Derrida proposes, there is nothing outside the text—"Il n'y a pas de hors-texte"—then a deconstructive method is open to a broad sense of how differential traces produce meaning: "The concept of text I propose is limited neither to the graphic, nor to the book, nor even to discourse. What I call 'text' implies all the structures called 'real,' 'economic,' 'historical,' socio-institutional, in short: all possible referents. . . . [E]very referent and all reality has the structure of a differential trace, and one cannot refer to this 'real' except in an interpretative experience."[62]

While I acknowledge that my many computer science interlocutors have been keen to tell me that an algorithm could never be a text, I consider the claim that the algorithm exists outside text to itself exercise a form of authorial power. Algorithms do certainly arrange differentially signifying traces to generate an output. Though the output may appear to have been generated directly, even autonomously, by a signifying code, those signs are also traces that are written into being to make the output present and presentable. One could seek out the source code of a deep neural net used in criminal justice, for example, but this code is categorically not the source from which the decisional outputs flow. As Derrida reflects on J. L. Austin's speech act theory, "He attempts to justify the preference he has shown in the analysis of performatives for the forms of the first person, the present indicative, the active voice."[63] The justification that Derrida finds so disquieting in Austin is the "source" of the first person from which the speech act is made. "This notion of source," Derrida suggests, is secured by "the presence of the author as a person who writes."[64] One could not locate a source, or a source code, since writing does not involve merely the transference of meaning from the author who writes to the world.

Understood in these terms, the algorithm used for genotyping or for risk targeting does not simply transfer the meaning contained within the source code into the world, for it has to write that into being. The deconstructive method offers a different kind of engagement with the text, one in which writing does not convey the message contained in the source but rather structures what can be said. To consider algorithms to be arrangements of differential traces (qua texts) is to open a different way to imagine the ethicopolitical contexts of the algorithm. Derrida speaks of the "non closure" and the "irreducible opening" of context, in which it becomes possible to find a "moment of strategies, of rhetorics, of ethics, and of politics."[65] Writing always exceeds its context as it travels, iterating and entering into new relations and new contexts. To conceive of algorithms in terms of the nonclosure of context is to resist rather directly the algorithm's determination to reduce the output signal to a single probability, to close the context. In the concluding pages, I sketch some of the parameters of an ethicopolitical engagement with the writing of algorithms.

The Unknowable Exerts Its Force

There are three broad parameters of the ethicopolitics of writing algorithms that I wish to draw together here. First, the nonclosure of context and the attention to how differential traces are arranged makes it possible to see more clearly how marginal or minor elements of a text might be functioning. Where one concept or meaning appears to dominate attention at the expense of a marginalized concept, a deconstructive reading is attuned to ask whether this minor concept might, in fact, be a crucial condition of possibility for the dominant meaning. Let me make this a little more concrete in terms of the marginalized or minor elements of the writing of algorithms. Consider, for example, the deep neural network whose sensitivity thresholds mark the boundaries of major/minor elements in the writing. Each layer in the algorithm transmits on to the next hidden layer only that signal it adjudicates to be "major" or to exceed a given threshold of probability. Though the input the algorithm receives is continuous, heterogeneous, and undifferentiated, the output signal has to be in the range 0–1; that is to say, it has to be expressed as a probability. To attend to the marginal elements of the writing of an algorithm is to return to the layers and to consider them points of nonclosure, where the making of a threshold of marginality is a necessary condition of the meaning of the dominant output. These points of nonclosure are moments not of illumination and transparency but of profound opacity, where what matters is the unknown, the undecidable, and the incalculable. This is not a trifling point of philosophical argument—it has real and palpable ethicopolitical implications. The DNA genotyping algorithms, for example, are reopened as text that is never completed and that always overflows its source code, so that what comes to matter is how the minor or marginalized data remain lodged within the layers of the algorithm. How is the likelihood ratio output of the algorithm transformed by moving the threshold between major and minor data elements? A strategy of sorts.

Second, the notion that there is nothing outside the text reorients critique so that one is less tempted to look outside the algorithm, to step outside its formation to find ethicopolitical grounds. Let us stay with the text, with the difficulties and undecidabilities of the text as such. A deconstructive strategy points also to how a single concept or word within the text contains within it different and irresolvable points of tension and argumentation. In the writing of algorithms, these concepts function simultaneously to advance the power of the algorithm to automate decisions and to interrupt and subvert them. Consider, for example, the use of *error* and *bias* in the writing of algorithms as models of decision. In the most evident public and political discussions of algorithms these concepts appear as sources of injustice and wrongdoing that must be removed to fix the algorithm's writing of the world. These same concepts, however, also have a quite different force that works to interrupt the idea that an algorithm could ever be something one might call unbiased or neutral. In short, the writing of the algorithm contains both the promise of an unbiased objectivity and the productive function of a bias that is required for the material to be worked on. In the writing of deep neural nets for speech recognition, for example, the algorithms "trained by back propagating error derivatives" are reported to have "the potential to learn much better models of data."[66] In this writing of the algorithm, weights and biases modulate the input signal to produce outputs as "guesses," where a feedback loop "rewards weights that support its correct guesses, and punishes weights that lead it to err."[67] The error derivative, just like the likelihood ratio in genotyping algorithms, is a generative fabulation that actively writes the world into being.

Finally, the differential traces of a text mean that it always already contains differences from itself. Thus, when patterns or styles are identified as being defining characteristics of a text—as with the algorithms predicting the next sentences of dead authors—those characteristics are undone by other signs carried within the same text. As Simon Critchley notes in his account of the ethical capacities of deconstruction, "Any text which identifies truth with presence as *logos*" will necessarily also enter into "contradiction with itself," signaling "towards a thinking that would be other to logocentrism."[68] For Critchley, this "hinge" between the closure of the logos and the opening onto the other is the very articulation of an ethics of deconstruction. Any demand for an enclosure of the context of writing, as for example in the many calls for the context of algorithmic systems to be defined and circumscribed, neglects this sense of the excess of context.[69] The idea that one could govern and limit the contextual application of a specific algorithm, for example, denies the constant iteration, modification, and adjustment of the writing of algorithms. Indeed, the algorithms used for extracting meaning from text, such as sentiment analysis, write themselves in and through decontextualized encounters, precisely by being exposed to something of the excess of context.

Following the writing of the algorithm involves watching for the moments when it enters into contradiction with itself. Computer science envisages the truth of the algorithm in presence or logos, where the algorithm adjusts itself to map signal to meaning correctly. In the writing, however, there are no definitive correlations but only continuous and open-ended iterations of the text, where every modification authorizes a new meaning and a new output. Indeed, this is what the idea of a likelihood ratio means in an algorithm used to determine the probability of the presence of someone's DNA at a crime scene. An infinitesimal modification in the training data (remembering that this could be exposure of the learning algorithm to the DNA data of racialized subpopulations), for example, will alter the meaning of the likelihood. At root, the output of the algorithm is not secured by the linear enunciation of its code. The person will appear differently as a suspect; another will be differently recognized as an immigration risk; another will become the optimal target for recommendations to vote or to shop: all of this because the iterative writing of the text gives rise to yet another set of inferred associations. Let us consider the writing of the algorithm, then, as though it were akin to Hilary Mantel's sense of fiction as showing the unknowable exerting its force.

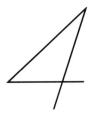

The Madness of Algorithms
Aberration and Unreasonable Acts

There are crises of reason in strange complicity with what the world calls crises of madness.
—Jacques Derrida, "Cogito and the History of Madness"

Reason demanded this step. In our science we have run up against the limits of what is knowable. . . . Our science has become a horror, our research dangerous, our knowledge lethal. All that is left for us physicists is to capitulate before reality.
—Friedrich Dürrenmatt, *The Physicists*

Errant Algorithms

There are moments when the everyday background hum of algorithms in society seems to break the surface and become something louder and more insistent. Such moments are often debated publicly as times when algorithms have become mad, or when they have departed from their rational logics and become frenzied, demented, and thoroughly unreasonable. In October 2016, when the British pound lost 6.1 percent of its value in a matter of seconds, the currency trades conducted autonomously by algorithms were reported to have precipitated a "flash crash," an eruption of irrational actions beyond the control of human traders.[1] Such apparent moments of the madness of algorithms, however, are not limited to the vicissitudes of the financial markets. In March 2016, Microsoft researchers experimented with a social media chatbot—named Tay and modeled on the attributes of a teenage girl—training her to communicate in a humanlike way by exposing her to the inputs of Twitter data streams. Within a few short hours, and in response to her Twitter learning milieu, Tay was posting racist and sexist messages, denying the holocaust and boasting to the police of her prolific drug habit.[2] Tay's deep learning algorithms were widely reported to be in meltdown, spiraling out of control in a descent into madness. The Microsoft experiment was finally halted when Tay messaged her two hundred thousand social media followers with "You are too fast, please take a rest," sending the message over and over again in a repetitious frenzy. The public's sense of dangerous algorithms becoming unreasonable and doing harm seemed to reach fever pitch in 2017, when YouTube's machine learning recommendation algorithms targeted automated content to children, playing videos with violent content on the YouTube Kids platform, such as the popular cartoon character Peppa Pig being tortured at the dentist, drinking bleach, and eating bacon.[3] In each of these moments, when the harms and dangers of algorithms seem to press themselves into public attention, what is most often discussed is a kind of algorithmic madness, a frenzied departure from reason.

As societies have responded to the errancy of algorithms, a specific ethico-political framing of the problem has emerged over time. Overwhelmingly, the public parlance around algorithmic ethics has placed emphasis on limiting the excesses of algorithms, controlling their impulses, and reining in their capacity for frenzied action in the world. The search for an encoded ethics for algorithms has tended to annex madness as an aberration that must be subjected to correction. In annexing the madness of algorithms as irrational aberration and departure from reason, such public interventions sustain the promise that algorithms could be rendered governable by appropriate thresholds of reasonable and unreasonable actions. Popular debates have depicted algorithms as "weapons of math destruction" that must be "taught right from wrong" in a world where one must "take care that ethical and not evil algorithms help win elections."[4] The propensity of the algorithm for madness, in short, has become a kind of limit point on its otherwise apparently reasonable place in our world. They can stay if they behave themselves. A kind of bounded rationality prevails, in which the boundary delineating good from evil in algorithmic decisions is one marked by reason versus unreason, rationality versus irrationality.[5] This boundary is breached, so it appears at least, when the algorithm acts in a way that was somehow not anticipated. Thus, for example, when the "kill switch" of a human intervention is activated—whether this is in an autonomous weapons system, artificial intelligence for speech, or a video recommendation system for children—this marks a threshold of something like madness, understood as a departure from reasoned logic.[6]

In this chapter I am concerned with how this common notion of the madness of algorithms as aberration has crucially overlooked the extent to which the rationality of algorithms is built on—indeed positively embraces and harnesses—the power of unreason. Such accounts of moral panic amid the madness of algorithms have had two significant effects. First, they have forgotten the place of notions of madness within histories of algorithmic rationality, where unreason is intrinsic to the computational logic. And second, they have underplayed the role of capricious incalculability within our twenty-first-century modes of algorithmic decision. Put simply, the common refrain of an encoded ethics for algorithms depends on a threshold that divides rational action from madness. The public discussion of the evils and harms of algorithms is focused on the regulation of that very threshold thought to separate rational from irrational action. Crucially, though, this threshold is, as all border lines are, the line that both pulls together and demarcates madness from all that is reasonable in the world. For Jacques Derrida, the division is a "dissension" or a "self dividing action," which both "links and separates reason and madness."[7] Discussing Foucault's treatment of the history of madness, Derrida cautions that "one cannot speak of madness except in relation to 'that other form of madness,' that is except in relation to reason."[8]

Taking seriously the conjoined histories of the ideas of reason and madness, I propose that one cannot speak of the madness of the algorithm except in relation to the form of reason the algorithm embodies. While the contemporary moral panic at each moment of the madness of algorithms urges us to police ever more vigilantly the line between reasonable and unreasonable actions, understood as a dissension, this line is precisely the condition of possibility of algorithmic rationality. Algorithms cannot be controlled via a limit point at the threshold of madness because the essence of their logic is to generate that threshold, to adapt and to modulate it over time. In short, my argument is that when algorithms appear to cross a threshold into madness, they in fact exhibit significant qualities of their form of rationality. Understood in this way, the appearance of a moment of madness is a valuable instant for ethicopolitics because this is a moment when algorithms give accounts of themselves. Contra the notion that transparency and the opening of the technological black box secures the good behavior of algorithms, the opacity and clouded action exhibited in the excesses and frenzies of algorithms yield a different kind of fidelity to the logic.

Let me make this concept of clouded action a little more concrete. Though the surfacing of violent images of beloved cartoon characters for preschool children appears as an undoubted algorithmic aberration, the computer science accounts of the development of deep neural networks for the YouTube video recommendation system exhibit a rationality that is consistent with outputting violent content. The algorithm designers describe the "dynamic corpus" of YouTube video content and explain their improved deep learning algorithms, which are "responsive enough to model newly uploaded content as well as the latest actions taken by the user."[9] Understood as a model that is trained on one billion parameters, the apparent madness of the recommendation of violent cartoons is explicitly part of, and not an aberration from, the rationality of the deep neural network. Indeed, the optimization of these algorithms is achieved in part through exposure to the new and emerging uploaded content, which is afforded greater weight in the model than user-specific data histories. At the level of the algorithm's logic—its mode of learning, internal weights, architecture, parameters, training data, and so on—the funneling of a frenzy of violent videos is entirely consistent with its rationality.

As I propose throughout this book, when viewed from the specific propositional arrangements of the algorithm, particular actions that might appear as errors or aberrations are in fact integral to the algorithm's form of being and intrinsic to its experimental and generative capacities. I am advocating that we think of algorithms as capable of generating unspeakable things precisely because they are geared to profit from uncertainty, or to output something that had not been spoken or anticipated. As machine learning algorithms increase their capacity to learn from raw unlabeled data streams, the unseen and unspoken become precisely the generative materials of algorithmic decisions. Such proximity between violent algorithmic outputs and the attributes of the data stream does mean that one has to think some heretical thoughts on ethics. The racist hate and misogyny of Tay; the funneling of far-right media to specific voter clusters using deep neural networks; the terrible consequences of image misrecognition in a drone strike—these are not instances when the algorithm has become crazed or frenzied. Rather, the learned action is a reasonable output given the extracted features of the data inputs. Of course, this is not a less troubling situation than the one in which some controls are sought on the worst excesses of the algorithm. On the contrary, it is all the more political, and all the more difficult, because that which could never be controlled—change, wagers, impulses, inference, intuition—becomes integral to the mode of reasoning. To be clear, my intention is not to avoid the need for critical response to the harms of algorithms. Far from it, my point is that violence and harm are not something that can be corrected out of an otherwise reasonable calculus.

In what follows in this chapter, I begin by resituating the genealogy of madness and reason in twentieth-century cybernetics, focusing specifically on the work of Norbert Wiener and the double enrollment of madness and rational forms of control into algorithmic systems. I then flesh out an alternative way of thinking about the madness of algorithms, reformulating the problem from one where algorithms might lose their hold on rationality to one where algorithms precisely require forms of unreason to function and to act. I am thereby amplifying attention to another kind of harm that does not reside beyond the threshold of something like the edge of reasoned action. This different kind of harm dwells in the algorithm's inability to ever embody the "madness of decision."[10] To live with the madness of decision is to acknowledge and take responsibility for the impossibility of ever binding the action to a full body of knowledge. Decisions are mad because they can never know fully the consequences and effects of their own making. To decide is to confront the impossibility of the resolution of difficulty; it is madness in the specific sense that it has no unified grounds.[11] With contemporary algorithms, decisions are being made at the limit of what could be known, and yet there is no responsibility for the unknown consequences of the decision. The madness of the decision is disavowed by the single output of the algorithm, and this disavowal is a potential horror and a danger. In the second half of the chapter, I discuss one specific set of algorithms—random decision forests—that have become the algorithms of choice across many domains, particularly in national security and border controls. The random decision forest algorithm grows multiple decision trees on random subsets of data and "takes a vote on their predictions."[12] It is an algorithmic proposition of What comes next? that takes place as a calculation amid incalculability, mobilizing chance and the splitting of agency, sometimes with lethal effects on human life.

Cybernetics and Unreason

In a 1950 essay, "Atomic Knowledge of Good and Evil," the cybernetician Norbert Wiener expresses his concern for the moral responsibility of science in the face of what he calls the "dangerous possibilities" of atomic weapons.[13] Reflecting on the nature of the roles of the mathematicians and physicists at Los Alamos, Wiener is anxious that "the new centralized atmosphere of the age to come is likely to lead society in the direction of 'Big Brother'" and toward a "future fascism." A full seventy years before scientists pointed to the dangers of a descent into fascism precipitated by the mathematical sciences of algorithms, the cyberneticians were writing publicly in newspapers and journals to express the view that "when the scientist becomes the arbiter of life and death" the locus of "moral responsibility" belongs properly to the reasoning scientist to prevent "the decay and disruption of society" and to ensure that science "is used for the best interests of humanity."[14] Indeed, in a 1947 letter to *Atlantic Monthly*, Wiener replies publicly to a request from government military officials for copies of his mathematics papers. Concerned that withdrawing his science from the service of war might be "shutting the door after the horse has become classified," Wiener believes that the locus of control for a dangerous science must reside in the reasoning human scientist. For Wiener, the rise of computing machines posed a threat defined not so much in terms of the madness of the autonomous machine as in terms of what he called the "crazed and misguided" irrationality of humans. "The automatic machine is not frightening because of any danger that it may achieve autonomous control over humanity," writes Wiener, but "its real danger" is that it may be "used by a human being to increase their control over the rest of the human race."[15] Wiener's letters and essays are replete with his anxiety that the science driving human progress may simultaneously be enrolled by sovereign or corporate powers and deployed for violence and war. For Wiener, the moral responsibility of the mathematical and physical sciences, particularly in their service of state power, resides in governing the line between reason and unreason. This distinction between reason and unreason, however, exposes something of the paradox of the place of madness in the cybernetic histories of algorithmic rationality.

First, the rise of cybernetics is closely intertwined with notions of overcoming the dangerous fallibilities of human judgment. Reimagining the human decision as a series of relays that could be modeled for optimization, Cold War cybernetics was geared to "tame the terrors of decisions too consequential to be left to human reason alone."[16] Understood in this way, the definiteness and conclusiveness of the algorithm as decision maker was thought to supply a rational safeguard against the madness of mutually assured destruction.[17] As Wiener wrote in his essay on "the new concept of the machine," the binary logic of computation structured decisions via a series of "switches," each offering a "decision between two alternatives" in a "logical machine."[18] For Wiener, the binary branching of the switch embodied what he called "two truth values," where the transmitted information was either "true or false." Here the logic of algorithms explicitly condenses the indefiniteness of data to a series of switches that must always be a binary choice of yes/no, true/false.

The branching that we see in classic decision tree algorithms, and in contemporary random forest algorithms, carries the traces of a cybernetic history, when the switch marked the calculation of truth and falsity. Amid the terrors of thermonuclear war, the promise of branching algorithmic decisions was that they could supply to political decision makers a means of avoiding the human propensity for frenzied miscalculation. At every instant of a branching decision, the logical machine rendered an output of truth or falsity. In this way, the twentieth-century rise of rational choice in foreign policy and behaviorism in the social sciences imagined a world in which algorithms limited the madness of a geopolitics on the edge of destruction. This first aspect of the place of unreason in genealogies of algorithmic decisions is significant to our contemporary moment because it is a reminder that algorithms are not devices that exist outside notions of the normative, norms, anomalies, and pathologies. On the contrary, the rise of algorithms in governing difficult and intractable state decisions makes them interior to defining what normalities and pathologies could be in a society. Just as the early neural network algorithms were embraced as a means of instilling reason into otherwise potentially dangerous human judgment, however, they also brought into being a new mode of harnessing the unknown and the unpredictable.

Second, at the same time as algorithms were thought to tame human irrationality, a specific orientation to madness was actively incorporated into the logics of algorithms. In Orit Halpern's book *Beautiful Data*, she gives a devastating account of how ideas about the psychotic and the neurotic were integral to cybernetic logics. "What has been erased from the historical record in our present," writes Halpern, "is the explicit recognition in the aftermath of World War II and the start of the Cold War that rationality was not reasonable."[19] In a detailed engagement with Warren McCulloch and Walter Pitts's development of the neural net, Halpern proposes that the logic of psychosis was crucial to the reformulation of computational rationality. Tracing the genealogy of neural nets as technical instantiations of the brain's functions, Halpern demonstrates the extent to which "neurotic, psychotic, and schizoid circuits proliferated in the diagrams of cybernetics."[20]

The insights of those who have studied the schizoid and paranoid relations of computation are critical to the contemporary ethicopolitical relations being forged with and through algorithms.[21] Far from representing a departure from rationality and entry into psychotic turmoil, the actions of algorithms are never far from their conjoined histories with psychosis, neurosis, trauma, and the imagination of the brain as a system. The double enfolding of unreason into cybernetic thought—as the unreasonableness of human decision makers in conditions of war and insecurity, and as models of psychosis underpinning the segmented cognition of the neural net—serves as something of a corrective to our contemporary societal debates on the responsibility of science for moments of algorithmic madness. Put simply, the history of algorithmic rationality is not separable from genealogies of madness. It is time to think difficult thoughts: algorithms are always already unreasonable. When one hears calls for new controls to *limit* the potentially dangerous actions of the algorithm, it is perhaps worth recalling that algorithms historically are *limit* devices. That is to say, they actively generate new forms of what it means to be normal or abnormal, just as they mark new boundaries of rationality and unreason. To capitalize on uncertainty—whether in warfare or in the commercial world of risk—algorithms dwell productively with emergent phenomena and incalculable surprises.[22] The actions of the contemporary neural net, for example, finds its condition of possibility in the twinned imperatives of limiting the impulses of human decision makers while embracing and harnessing the impulses of models of the brain as nervous circuits.

To be clear, contemporary algorithms are oriented toward limiting the unanticipated actions of humans while generating a whole new world of their own unanticipated and unreasonable actions. This means, for example, that an algorithm will decide the threshold of whether a person's presence on a city street is normal or anomalous, and yet it will generate this threshold via experimental methods that enroll errancy and capricious action into the form of reason. In the section that follows, I revisit the problem of madness to reformulate the algorithm's relation to unreason.

Madness and Classification

Throughout this book I urge some caution about the treatment of the ethical and political questions posed by algorithms as novel or unprecedented. Indeed, I suggest that the advent of algorithms as decision makers in our societies has restated what are in fact perennial philosophical debates about the nature of ethical and political life. Among these perennial debates is the question of whether and how one controls the impulses or actions of an entity that appears to us as beyond human reason. It is in the philosophies of human madness and its relationship to forms of reason, I propose, that one can locate a resource for thinking differently about the actions of algorithms. In Michel Foucault's major genealogical work on the history of madness, he argues that madness is designated as a "blatant aberration" against all that would otherwise be "reasonable, ordered and morally wise."[23] Understood in this way, the very idea of human reason has required—integral to its historical condition of possibility within a moral order—an opposition to the aberration of madness. Indeed, for Foucault, "in our culture there can be no reason without madness," and yet the objective science that observes, records, and treats pathologies of madness simultaneously "reduces it and disarms it by lending it the slender status of pathological accident."[24] Thus, there is a necessity of madness within Western thought, and yet this is a diminished form of madness reduced to a pathological accident or error. The idea of unreason is always already present in claims to moral reason, but this is an unreason understood as a flaw or error in the logic of reason—a flaw that can be identified, classified, and cured.

Such a sense of a reduced form of madness within Western moral order seems to me to be amplified in our contemporary moment, when the actions of algorithms are so frequently named as accidents or errors. The threshold of madness denotes a boundary between the reasonable and the unreasonable, just as it also condenses the unknowability of madness to an error, an accident, or a flaw in the code. Significantly, Foucault's history of madness traces what he calls "a vast movement" from critical to analytical recognition of madness, where classical forms were "open to all that was infinitely reversible" between reason and madness, while the nineteenth and twentieth centuries brought the "classificatory mania" of psychiatry and insisted on new lines demarcating the rational from the mad.[25] The opposition of madness and reason, then, is a historically contingent event that occludes and reduces a more expansive sense of the human experience of madness.[26] As Jacques Derrida describes it, "The issue is to reach the point at which the dialogue" between reason and madness is "broken off" and to permit once more a "free circulation and exchange" between reason and unreason.[27] The modern break in the dialogue between reason and madness, described by Derrida as a "dissension," makes it necessary, he proposes, to "exhume the ground upon which the decisive act linking and separating madness and reason obscurely took root."[28]

The dissension linking and separating madness and reason is a relation rather different in character from the limit point of a madness as aberration from reason. Rather, the dissension itself makes possible new ways for society to understand and to govern the relation between reason and madness. It is an ethicopolitical dissension that generates not only new models of what is normal and abnormal in human societies, but crucially also new ethical relations between selves and others.[29] The relationship between madness, science, and law from the nineteenth century became a moral force in which the rationality of the psychological sciences would assert the boundaries of the normal and the pathological. With the advent of positive psychiatry, as Foucault describes, a "new relation" became possible between the condition of madness and "those who identified it, guarded and judged it" within the "order of an objective gaze."[30] Hence, the dissension forges ethical relations that render madness knowable "at a stroke," authorizing "reasonable men to judge and divide up different kinds of madness according to the new forms of morality," just as it simultaneously tames the nonknowledge of the experience of madness.[31] In the making of a particular ethical and moral relation to madness, Foucault locates a first major objectification of the human:

> From this point onwards, madness was something other than an object to be feared. . . . It became an object. But one with a quite singular status. In the very movement that objectified it, it became the first objectifying form, and the means by which man could have an objective hold on himself. In earlier times, madness had signified a vertiginous moment of dazzlement, the instant in which, being too bright, a light began to darken. Now that it was a thing exposed to knowledge . . . it operated as a great structure of transparency. This is not to say that knowledge entirely clarified it, but that starting from madness a man could become entirely transparent to scientific investigation.[32]

To make an object of madness, then, was to make a world of transparency in which an individual could have a hold on himself. Let us consider this notion in light of our contemporary moment, when once more madness is becoming object in terms of both the possibilities of rational objective algorithmic decisions and the demand for algorithms to take a hold of themselves and to be rendered transparent to investigation. The moment of dazzlement and darkening Foucault describes, when the place of madness within reason is acknowledged, is entirely absent in the contemporary calls for transparency and the opening of the black box of the algorithm. I wish to reopen the dissension in madness and reason that Derrida and Foucault differently depict and make of this dissension a different kind of ethical terrain. With dissension conjoining madness and reason, the ethical move is no longer a matter of the search for moral codes that regulate the boundary between rational and unreasonable behavior by algorithms. Instead, the manifestation of a moment of apparent madness in an algorithm's actions becomes necessarily also a moment when the algorithm gives an account of its form of reason. Rather than contribute to the cacophony of calls for greater transparency, this ethical demand dwells with the vertiginous moment of dazzlement and darkness, with clouded action and opacity, so that the violences of an algorithm cannot simply be mistakes or errors in the algorithm's otherwise logical rationality.

When one no longer seeks transparency and a hold on oneself, the appearance of a moment of algorithmic madness offers a different kind of insight—an insight into how the algorithm enrolls and deploys ideas of unreason to function and to act. In the moment of madness, we see a dazzling instance of the form of rationality's improbability. In turn, the question of responsibility also shifts ground. Reflecting on the rise of positive psychiatry, Foucault notes that unreason retained a moral dimension, in which "madness was still haunted by an ethical view of unreason, and the scandal of its animal nature."[33] The responsible subject, then, would act to annex the animal from the human and to locate ethics in the unified mind. In his discussion of homicidal mania and responsibility for murder, for example, Foucault notes that for the defendant to be responsible for the act, there had to be "continuity between him and his gesture."[34]

To be mad in the sense of being outside oneself was to be caught and "alienated," divided within oneself so that a person "was himself and something other than himself."[35] This dividuated subject is precisely embraced by the actions of the algorithm so that its posthuman form is to be simultaneously human and something other than human, a form of self, and fragments of the other. So, responsibility cannot feasibly take the form of taking a hold on algorithms or annexing instinctive or impulsive animal behaviors. I want to take seriously the idea that "in the absence of a fixed point of reference, madness could equally be reason," so that there can be discontinuity between an algorithm and its actions while a responsibility still remains.[36]

To reverse the opposition of madness and reason would be to think radically differently about the many moments when algorithms have been said to err or to deviate from accepted norms. The logic of algorithmic errancy expresses the responsibility of the algorithm precisely in the terms of there being continuity between the algorithm and the gesture. When the algorithm strays or deviates, it is considered irresponsible because it no longer controls the outputs it effects. And yet, as I argue in previous chapters, the experimental straying from paths defines much of the power of contemporary machine learning. A kind of gestural discontinuity, in short, is profoundly useful to the algorithm. To reduce madness to error or aberration is to "neutralize madness" and to shelter "the Cogito and everything related to the intellect and reason from madness."[37] The errors and aberrations of algorithms continue to dominate public discussion of the ethics of machine learning, automated decisions, and data analytics. This dominance of errors and aberrations is indeed neutralizing madness in its broadest sense—as that which cannot be fully known or spoken—and sheltering the rationality of algorithms from a full and expansive critique. The illegible, unspeakable, and opaque actions of algorithms, as I argue throughout this book, are not the limit points of what is possible ethically; instead they are the starting points of an ethical demand.

When the effects of algorithmic decisions are truly horrifying—such as in Cambridge Analytica's rendering of the attributes of "persuadable" voters in the US presidential election and the UK EU referendum (with its use of the attribute to target xenophobic anti-immigration media, for example)—reining in their crazed excesses can never be sufficient. The unreason and the excess are not aberrations at all but are the condition of possibility of action. The problem seems to be, then, not that the rational and the reasonable algorithm takes leave of its senses, loses control, and loosens its hold on its logic. Rather, the algorithm is always already beside itself and divided within itself as such.[38] Algorithms learn by unrestrained experimentation and emergent signals. They simultaneously send multiple conflicting signals along different pathways to optimize their output. They are errant not in the strict sense of deviating from a path, but in the archaic literary sense of traveling in search of adventure. And so, algorithms must also be understood differently in their actions—not as entities whose propensity for madness can be tamed with correct diagnosis and repair, but instead as entities whose particular form of experimental and adventurous rationality incorporates unreason in an intractable and productive knot.

I examine this generative mode of unreason as it animates the decisions of algorithms dealing with human life. The violences and injustices that result from the algorithm's decisions do not emerge primarily from errors, accidents, or aberrations in the system's logic. Such a framing of the violent outputs as errors shelters the cogito from the darkness of unreason. It also neutralizes the fullness of madness as improbability and the unknowable, thus restricting what can count as the harms of algorithms. A principal harm of algorithms is that they enable calculative action where there is incalculability and the unknowable, reducing ethicopolitical orientations to the optimization of outputs and the resolution of difficulties. Renewed attention to the unreason within the algorithm animates how combinations of probabilities generate improbable and untamed outputs that are let loose into the world.

The Madness of Decision

In October 2017, the US National Science Foundation (NSF) awarded a $556,650 research grant to a team of engineers and philosophers working on the decision-making algorithms for autonomous vehicles.[39] The research represents a rather direct example of what I have termed an *encoded ethics*, in which codes of conduct are sought to modify and restrain the harmful effects of algorithmic decisions. The research addresses the classic moral philosophy "trolley problem," redefined for the age of algorithms. This is a scenario in which a trolley car is careering down the tracks toward a group of five people. The driver faces a profoundly difficult decision—to continue on the track toward the certain deaths of the five people or to pull the lever, thereby intentionally and fatally redirecting the trolley onto a second track where one person lies immobilized. In moral philosophy, the trolley problem is intended to highlight the problem of the grounds of rational decision making. How should the driver weight the value of the lives on the tracks? Is this calculation morally questionable? Should she make an active decision to kill one person to avoid the passive accidental deaths of five? In the NSF-funded research, the trolley problem is reinterpreted for decisions made by algorithms guiding autonomous vehicles. "You could program a car to minimize the number of deaths or life-years lost in any situation," explain the researchers, "but then something counterintuitive happens: when there's a choice between a two-person car and you alone in your self-driving car, the result would be to run you off the road."[40] The research imagines a world in which algorithms can be trained to precompute the rational decision in the face of a scenario of catastrophic consequences.

The boundary of this rationality is a computational threshold in which the algorithms weigh the values of different choices before making a decision. The research team is working on deep neural networks that will "sort through thousands of possible scenarios," filtering out and "rapidly discarding 99.9% of them to arrive at a solution." The example the scientists discuss is a self-driving car hurtling toward a school bus, where the optimal pathway is thought to be "discarding all options that would harm its own passenger" before "sorting through the remaining options to find one that causes least harm to the school bus and its occupants."[41] Here, in the starkest of terms, is a statement of the kind of ethical codes sought by many engineers and algorithm designers.[42] It is an ethical mode that promises to render calculable the profoundly uncertain horizon of an immediate future. Put simply, in seeking to ward off the full horrors of frenzied algorithms governing a trolley out of control, such programs generate new harms in the valuing and weighting of different pathways. The calculus itself—with its combinatorial possibilities of babies in strollers, elderly pedestrians crossing the road, careless delivery drivers, and inattentive cyclists—is always already thoroughly unreasonable. Algorithms do not pose their most serious and harmful threats when they are out of control, become crazed, or depart rationality. On the contrary, among the most significant harms of contemporary decision-making algorithms is that they deny and disavow the madness that haunts all decisions. To be responsible, a decision must be made in recognition that its full effects and consequences cannot be known in advance. A responsible decision-making process could never simply "sort through the remaining options to find one that causes the least harm" because this is an economy of harms that renders all the incalculability of harm calculable.

The madness of the trolley decision is that it must necessarily be made in the darkness of nonknowledge, that it categorically is not subject to pre-programming to optimize an outcome. As Derrida writes, "Saying that a responsible decision must be taken on the basis of knowledge seems to define the condition of possibility of responsibility," and yet at the same time, "if decision-making is relegated to a knowledge that it is content to follow, then it is no more a responsible decision, it is the technical deployment of a cognitive apparatus, the simple mechanistic deployment of a theorem."[43] Though the algorithm's design for the trolley problem may appear to be the simple mechanistic deployment of a theorem, it does contain the aporia of multiple decisions. The problem is that these multiple, fully ethicopolitical decisions are themselves annexed and sheltered from madness by means of errancy or mistake.

One route into the fuller ethicopolitics of an algorithmic decision is to attend to its vast multiplicity. Thus, as I explain in chapter 2, the autonomous vehicle is unable to recognize "child," "stroller," or "school bus" without a regime of recognition trained on millions of parameters of labeled images. The computer scientists working on encoded ethics for the trolley problem (at the heart of the ethicopolitics of all autonomous vehicles) seriously underestimate the difficulty when they speak of filtering "thousands" of possible scenarios, for each of these scenarios is nested within millions of other parameters. An algorithm cannot be programmed to value, for example, the life of a child or an adult as such, for even this decision contains within it the multiple fraught difficulties of learning how to recognize or misrecognize a person. Where the introduction of moral philosophy into engineering problems has sought to contain the algorithm's capacity for an irrational output, the algorithm in fact positively embraces and requires irregularities, chance encounters, even the apparently errant past actions of humans to learn how to optimize the output. The common experimental method of "spoofing" an algorithm in development, for example, involves the exposure of the algorithm to a kind of frenzy of false or "spoof" images to teach it to refine its capacity to recognize truth from falsity. To summarize at this point, when an autonomous vehicle appears to have departed from its logic, lost control, or become mad, it has in fact yielded to the world something of how its logic functions, how its decisions are arrived at. Rather than seeking *an encoded ethics* to try to limit the madness of algorithms, *a cloud ethics* should proceed from the incompleteness and undecidability of all forms of decision. The algorithm contains within its arrangements all the many multiples of past human and machine decisions. It has learned how to learn based on the madness of all decision, on the basis of nonknowledge. To have responsibility for decisions on life, as the autonomous vehicles' algorithms manifestly do, is to foster conditions in which that learning requires unreasonable things. In the sections that follow, I explore two sets of circumstances when algorithms are deciding on life in the face of an incalculable future.

Life and Unreason I: "Exaggerated Results"

In the summer of 2016 a group of Swedish computer scientists published their findings on algorithms designed to interpret images of the brain from functional magnetic resonance imaging (fMRI) scans. Their investigations into the statistical validity of the major algorithms used globally in neuroimaging seemed to show a stark and troubling rate of error, what they called a "cluster failure."[44] In the context of a method used for neuroimaging over a period of twenty years, the finding that the algorithms have a 70 percent false positive rate cast fundamental doubt on scientific knowledge used in important areas such as Alzheimer's research.[45] A 70 percent false positive rate may appear to be an errant departure from the statistical logics governing algorithmic inference. The neural networks did not depart from their logic, however, but generated what were described as "exaggerated results" in and through that very logic. The deep neural networks were identifying clusters of brain activity in 70 percent of cases where there was no actual cerebral activity present. Crucially, though, the algorithms had been exposed to millions of training images of brain activity to recognize the attributes of new and unseen instances. Here, once again, there is a regime of recognition generated through the algorithm's exposure to the features of past datasets. Among the many assumptions dwelling within the algorithms was a set of norms about the morphology of the brain that ultimately led to misrecognition and a high false positive rate. In short, the proliferation of false positives, with serious consequences for the neurological diagnoses of countless people, was not an aberration in the algorithm's rationality but in fact a manifestation of the excesses of its form of reason.

When multiple false positives are produced in other domains of algorithmic decision, such as in facial recognition systems, often these are understood as errors that could be corrected out.[46] Yet, if one considers that to reduce madness to error is to protect reason, then the resort to error does not limit the algorithm's use in society but actually helps it to proliferate into new domains of life. In my reformulation of the madness of algorithms, I wish to revalue the uncertain relationship to truth that is embodied in algorithms. "Madness begins where man's relationship to the truth becomes cloudy and unclear," writes Foucault, so that to experience madness is to experience the "constant error" of the destruction of one's relation to truth.[47] To correct the error would be to seek a kind of transparency and clear-sightedness that would create a break in the cloudiness. Yet, if one stays with the cloudiness of nonknowledge, then a different kind of partial and occluded account is demanded.

What could this different calling to account and responsibility look like? With a cloud ethics, where one is proceeding with the partiality and opacity of all forms of accounting for oneself and others, one would seek to show how and why algorithms can never bear responsibility for undecidability, or for the madness of all decision. For example, when Pedro Domingos makes his bold claims that machine learning algorithms are "learning to cure cancer" by analyzing "the cancer's genome, the patient's genome and medical history" and simulating "the effect of a specific patient's mutations, as well as the effect of different combinations of drugs," a cloud ethics would acknowledge the multiple and distributed selves and others dwelling within the calculus, signaling the madness of the decision itself.[48] The rapid increase in the use of algorithms for genotyping and cancer treatment decisions is marked by the gathering of sufficient data parameters to produce optimized treatment pathways as outputs. But, of course, in the context of health care, there is pressure to include economies within the parameters of these algorithms, such as the cost of the drugs, the benefit to the pharmaceutical industry, the likelihood of "quality life years," or the time efficiency to the overstretched clinician.[49] While the responsible decision of an oncologist is made in a way that is cognizant of the unknowability of the outcomes of rejected treatment pathways, the algorithm's decision proposes the optimal pathway on the calculative basis of the weightings of all possible pathways. Such processes of precomputational decision can never bear the full weight of the madness of all decision. In effect, the precomputed weighting of different branching pathways we saw in the trolley problem is mirrored in the different algorithmic pathways of cancer treatment. In both cases—and there are many more—where the outcome will certainly be the loss of life for some and the preservation of life for others, the madness of the decision as such is disavowed. The madness of the algorithm does not reside in the moral failure of the designer or the statistician, but it is an expression of the forms of unreason folded into a calculative rationality, reducing the multiplicity of potentials to one output. The madness of the algorithm expresses something of what cannot be said; it is the absence of an oeuvre, gesturing to that which is inexplicable and unspeakable.[50]

Life and Unreason II: SKYNET and the Random Forest

In the preceding section, where I discuss a 70 percent false positive rate in neuroimaging algorithms, the matter of life and death hinged, at least in part, on the capacity of the algorithm to recognize the singularity of the patient's brain as the terrain of a target. In other places where the deep neural net algorithm travels, the matter of harm can similarly hinge on the recognition of a target, though in ways in which even a relatively low false positive rate can have catastrophic consequences. One specific kind of algorithm, the random forest, or *random decision forest*, has become the algorithm of choice for systems designed to identify terror targets for the state.[51] Why might this be? Because machine learning for counter-terrorism is a difficult problem. Unlike the commercial algorithms for credit card fraud or email spam detection, for example, where there is vast availability of labeled training data on which to train the classifier algorithms, the availability of labeled data on known terrorists is extremely scarce, indeed even thoroughly inadequate for statistical modeling. This has meant that terrorist-targeting algorithms have become, in a rather direct sense, *unreasonable*. That is to say, in the context of profoundly uncertain security futures, these algorithms have harnessed the unknown and the incalculable to preserve the capacity to act. I am going to dwell a little longer on the random forest algorithm as an arrangement of propositions in the world. The randomness of this algorithm is, I suggest, a kind of sheltered madness that dwells inside the logic of the algorithm and promises to the world an impossible vision of a secure future. It is, in short, an algorithm with life and death effects, and it generates these effects by reformulating the dissension between reason and madness.

In October 2016, two journalists reported that the NSA's SKYNET program "may be killing thousands of innocent people."[52] In the context of the estimated four thousand people killed by drone strikes in Pakistan between 2005 and 2016, Christian Grothoff and Jens Porup investigated how SKYNET "collects metadata," storing it "on NSA cloud servers" and then applying "machine learning to identify leads for a targeted campaign." Among the "cloud analytic building blocks" described within SKYNET in the Snowden files are travel patterns; behavior-based analytics, such as incoming calls to cell phones; and other "enrichments," such as "travel on particular days of the week," "co-travelers," and "common contacts" (figure 4.1).[53]

I have elsewhere written in detail on how these aggregated security "data derivatives" are generated and how they are made actionable.[54] Of specific interest here are the random forest decision tree algorithms that are learning in

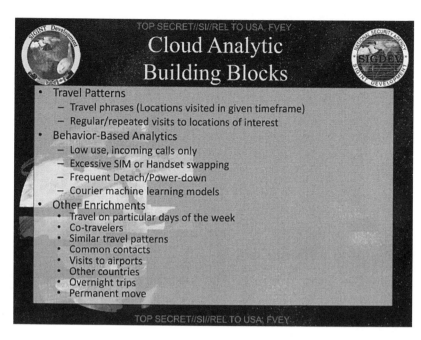

Figure 4.1   SKYNET cloud analytic. Cryptome.org.

communion with the data derivatives. Random forest algorithms were developed at the turn of the millennium by Leo Breiman of UC Berkeley's Department of Statistics.[55] These machine learning algorithms exemplify well how notions of chance, preference, error, and bias become incorporated into the capacity to compute an output. As Breiman explains, in the random forest technique, the algorithms come into being by "growing an ensemble of [decision] trees and letting them vote for the most popular class."[56] Noting that "data sets with many weak inputs are becoming more common" (for example, in medical diagnostics) and that "this type of data is difficult for the usual classifiers—neural nets and trees"—Breiman designed his algorithms to "inject the right amount of randomness" so that the algorithm is trained on random subsets of the training data and "predicts the class of each new record using the plurality of the class predictions from the set of trees."[57] In this way, the random forest algorithm actively uses variance and randomness to refine its output. As Lorraine Daston has argued in her compelling account of the histories of probability, chance, and unreason, all three were able to flourish within methods of calculation.[58] Contemporary algorithms such as random forests have found new ways to invite randomness and chance back into the probabilistic calculus.

If the random forest algorithm appears to descend into the horror of targeting civilians for drone strikes in the SKYNET program (and I have also seen it used by governments to output a risk score of immigration infraction), how can one understand the mode of unreason that dwells within the form of algorithmic rationality? A key element of the random forest algorithm's logic is the attributes-based targeting of population I discuss in previous chapters. The SKYNET random forests are trained on data from a population described as "known couriers." These proxies for labeled data on terrorists are individuals within Pakistan who have been scraped from intelligence reports under the keyword "courier," denoting an individual who is suspected of carrying messages for known terror groups, such as al-Qaeda. The random forest algorithm learns the attributes of "indicative behaviors" from the training data of couriers and then runs the data of 55 million mobile phone users in the general population of Pakistan through this model to "discover similar selectors with courier-like travel patterns."[59] As the random forest algorithm is increasingly now applied to a mobile data stream (and not only a static dataset)—for example, in the video feed of an unmanned aerial vehicle (UAV)—this discovery of similar behavioral patterns is taking place without human input, so that the algorithm is said to "tune its own parameters in response to the data it sees, relieving the analyst of the need to carefully define algorithm parameters."[60]

Though random forest algorithms could be said to generate a kind of madness of false positives that become actionable as a kill list, I am proposing that in fact this madness is useful to the algorithm. Unreason supplies a refinement of the algorithm's logic so that it is no longer limited by the gaze of the analyst. SKYNET's false positive rate is 0.008 percent, where a 50 percent "miss rate" (the false negative rate, or where half of those with "courier-like" patterns are missed) is tolerated in the model. To signal this as error or mistake, however, or to try to fix the problems in the training and validation data is manifestly to overlook the place of unreason in the algorithm's learning. The SKYNET random forests generated one high-profile false positive—recognizing the travel patterns of the Islamabad bureau chief of Al-Jazeera, Ahmad Zaidan, as a high-scoring selector—that was represented as errancy or mistake. To be clear, according to the logic of the random forest, the recognition of Zaidan is not errant at all but is precisely a rational selector given the proxy patterns of the "couriers" and the class predictions of the set of trees. The classifier is perceiving a scene through the apertures of the branches and leaves of the random forest. It learns to distinguish things in the world precisely through the injection of randomness. In the security and defense domains, this capacity of the random forest algorithm to distinguish through variability is highly valued:

"As the classifier watches events unfold, it tries to discern patterns of behaviour: a pack of wolves circling a wounded animal, shoppers taking items from store shelves to a cash register, or an insurgent burying an IED on the side of the road. It's the algorithm's job to learn how to distinguish between someone just digging a hole and someone else burying a bomb."[61]

For the random forest algorithm, the unreasonable and untamed wildness of the random becomes the means of recognizing and distinguishing a target. The random element and the incalculable are lodged within its form of reasoning. To signal the importance of this enfolded unreason is not to deny the profoundly violent effects of deploying the random forest algorithm for terrorism targeting. Far from it. In fact, attention to the unreasonableness of the algorithm as such highlights the impossibility of confining or constraining the madness of the algorithm. If the civilian digging a hole in Pakistan, or the school bus at the border, is mistakenly targeted, the madness of the drone strike is sheltered by the generation of new input data that modifies the threshold between "false alarm" and "miss rate." Every output of a target, however error prone or crazed in its assumptions, supplies new input material to the model.

Given what we know about the place of psychosis and schizoid circuits in the cybernetic history of algorithms, and about the genealogies of the relationship between madness and reason, perhaps it should not be surprising that the twenty-first century is witnessing a reformulation of unreason within the rationality of algorithms. Nineteenth-century diagnosis and treatment of madness and dementia reimagined madness so that it "was not an abstract loss of reason" but instead "a contradiction in the reason that remained," rendering the patient's rationality recoverable.[62] In many ways this notion of a flaw in reason that is repairable has dominated discussion of the public ethics of algorithms. Yet, the modern notion of madness as an error in reason is undergoing reformulation with the advent of decisions involving human and algorithm collaborations. The twenty-first-century violences of algorithmic logics point not to contradictions or flaws in reason that could be cured by the rational human in the loop, but to the potentialities of the excess of reason and the power of unreason. In algorithms such as random forest, the demented wager, randomness, and chance become newly reacquainted with rationality.[63] To treat the pathologies of algorithms, then, is simultaneously to engage in a "forgetting of violence and desires" that continues to animate the algorithms arbitrating the threshold between life and death in our times.[64] A random forest algorithm will never know a terrorist in the sense of acting with clear-sighted knowledge, but it mobilizes proxies, attaches clusters of attributes, and infers behaviors to target and to act regardless.

The Scaffold: Death Penalties and the Condition of Not Knowing

The madness of algorithms, as I have reinterpreted the condition in this chapter, arises not from the loss of rationality or from the error-prone deferral of human reason into the machine, but instead from the making of a calculation in conditions of nonknowledge.[65] When an algorithm generates a singular output from an incalculable multiplicity of associative relations, it shelters the darkness of the decision in the reduction to one actionable output. In Jacques Derrida's writing on the death penalty, he describes "the imposition of *calculability* on a condition of non-knowledge." For Derrida, the specific violence of the death penalty (even as it compares to other modes of killing) is that an unknowable future—"the given moment of my death"—becomes calculable "with absolute precision." "Who thus calculates," writes Derrida, "turns justice into a utilitarian calculation" and makes the death of a person into a "trade, a useful transaction."[66] My argument is that contemporary algorithms are extending this useful transaction of the programmable decision—the useful output, the good enough model—into other calculations at the threshold of life.

The weights, thresholds, and attributes through which an algorithm comes into being are simultaneously the condition for assigning the calculative weight or the value of someone or something.[67] Just as Derrida identified in the US penal architecture that "the majority of those condemned to death are blacks and poor blacks," our contemporary times witness a calculating machine that also generates weighted and racialized targets and sentences. Despite the manifest differences between the discrete logics of particular machine learning algorithms, what they all share in common is the reduction of a multiplicity of incalculable differences to a single output. This output is a finite target contingent on infinite multiples of weights and weightings within the calculation. Thus, when a random forest algorithm sentences someone to death by drone strike, the infinite (gestures, connections, potentials) makes itself finite (optimal output, selector, score), and the horizon of potentials is reduced to one condensed output signal.

Moreover, this reduction to one output signal, a kind of death penalty in its making of precise action from nonknowledge, shelters itself from its own unreason. As societies seek out an encoded ethics that annexes and extracts the errant, the aberrant, from algorithms, they are failing to understand how these apparent pathologies are actually of the essence of algorithmic learning. The clustering of false positives in a particular black population in one part of a city, for example, if understood as errant becomes subject to correction and recalibration. Yet, this experimentation and wild errancy continues to generate targets long after a correction is made. The random forest algorithms in SKYNET, as well as those in immigration systems that return targeted migrants, condemn to death unknown people living and traveling in already risky spaces, and they do so with wagers, votes, chance, and randomness. The algorithm underwrites its own rationality because it is engaged in defining new thresholds of normal and abnormal behaviors, reasonable and unreasonable travel, the arbitration of the good and the bad.

The architecture of algorithms adjudicating life and death takes the form of a kind of scaffold, where the scaffold is the spatial arrangement of the death penalty, the place of execution.[68] In this sense the algorithm as scaffold is part of the scaffolding of sovereignty, where the state "will have constructed all the scaffolds and propped up all the figures of machines for killing legally, sovereignly, nationally in the history of humanity."[69] The algorithm as scaffold is sometimes—as with random forest algorithms for national security—the means by which states exercise the right to precisely calculate death in advance; but it is also the architecture of a decision, a "certain modality, a certain qualification of living and dying, a manner, an apparatus, a theatre, a scene of giving life and giving death." Moreover, the scaffold "guarantees some anonymity" for the executioner, who is not present on the scaffold and whose hand is not on the guillotine, an "executioner who does not kill, not in his own name."[70] The random forest algorithms I discuss in this chapter, as scaffolded arrangements, similarly apportion weights to life and death as output signals, guaranteeing some anonymity to the executioner, who does not appear on the platform. The algorithm is precisely a space in which the excesses, the unreason, the cruelty of the adjudicator can be given free rein. And so, to find a critical response to the algorithm via a demand for limits on its unreasonable excess is to seriously overlook how exactly it deploys unreason to generate a finite output.

Part 3

Ethics

The Doubtful Algorithm
Ground Truth and Partial Accounts

During the war I worked on the atomic bomb. This result of science was obviously a very serious matter, it represented the destruction of people. . . . Is there some evil involved in science? Put another way—what is the value of the science I had dedicated myself to—the thing I loved—when I saw what terrible things it could do?
    —Richard Feynman, *What Do You Care What Other People Think?*

I remain a child of the Scientific revolution, the Enlightenment, and technoscience. My modest witness cannot ever be simply oppositional.
    —Donna Haraway, *Modest_Witness@Second_Millennium*

Leave Room for Doubt

In a lecture delivered to the National Academy of Sciences in 1955, physicist Richard Feynman reflected on a particular relationship between the practice of doing science and the value of a kind of orientation to *doubt*. As a graduate student in the 1940s, he worked on the physics of the atomic bomb at Los Alamos. Later, observing the terrible consequences of the weapons to which his physics had contributed, he began to doubt the value of science and its responsibility to society.[1] In the years that followed his work on the Manhattan Project, Feynman locates the practice of science in a particular method that permits a freedom to doubt, an animated curiosity for otherwise ordinary things, and a sustained sense of encountering an unknown future world. In his 1955 lecture, he describes how the scientist "must leave room for doubt"; proposes that "it is perfectly consistent to be unsure, it is possible to live and not know"; and makes a claim about responsibility to society: "Permit us to question, to doubt, to not be sure, herein lies a responsibility to society." For Feynman there is a profound ambivalence within the idea of doubt, so that it expresses simultaneously "the openness of possibilities" into the future *and* the doubt-fulness of science in the service of the state's "monstrosities" of war.[2]

Feynman's reflections on the responsibility that twentieth-century science has for violence and war testify to the intractable difficulties of delineating ethical lines of good and evil in technoscience. Along with many other physicists whose experimental fragments became lodged within a horrifying calculus, Richard Feynman embodies what Rosi Braidotti has called the "anti-Humanism" of the "turbulent years after the Second World War," when the profound tyranny of human rationality was felt.[3] Yet, as Donna Haraway acknowledges of her own subjectivity as "a child of the Scientific revolution, the Enlightenment, and techno-science," what it means to be human is not separable from technoscience and its place in war.[4] Though Feynman's situated and partial accounts of the science of the atomic bomb bear witness to the state's enrollment of algorithmic calculative practices in war, these are not accounts that stand outside science in opposition. Rather, the partial account dwells uneasily within the interconnected narratives, stories, writings, and narratives that together constitute the practice of science.

What is striking to me, in the letters and diaries of twentieth-century mathematicians and physicists, is that the intransigent and political difficulties of partial accounts—manifest in the public debates of the twentieth century—are all but entirely absent in the twenty-first century's enrollment of algorithms by the state. In the demand for transparency and clear-sighted accountability, the algorithm's orientation to doubt is never in question. Amid the pervasive twenty-first-century political desire to incorporate all doubts into calculation, algorithms are functioning today with the grain of doubt, allowing the uncertainties and variabilities of data to become the condition of possibility of learning and making decisions. In contrast to Feynman's notion that science's responsibility to society may reside in leaving open the incalculability of the future, algorithms hold out the promise of securing against all possible future events. With contemporary machine learning algorithms, doubt becomes transformed into a malleable arrangement of weighted probabilities. Though this arrangement of probabilities contains within it a multiplicity of doubts in the model, the algorithm nonetheless condenses this multiplicity to a single output. A decision is placed beyond doubt.

As doubt becomes the terrain of algorithmic calculation, however, other situated and embodied forms of doubt haunt the terrain. These opaque and cloudy accounts seem present at all the moments when an algorithm generates itself in relation to a world of data. For example, when a biometric matching algorithm doubts it can recognize a face because the data points were not present in the training data, or when four desk analysts watching luminous screens of risk scores worry that the model is "overfitting" to the data—these are moments not of a lack or an error, but of a teeming plenitude of doubtfulness. What might it mean to invoke Feynman's "room for doubt" or "to live and not to know" in the context of a technoscience driven to expose machine learning algorithms to data precisely to know and to act, indifferent to persistent doubt?[5] In light of the algorithm's major role in the calculability of doubts, in this chapter I propose that a reformulated doubt opens up new potentials for cloud ethics. The post-Cartesian and posthuman form of doubt I envisage begins from accounts of doubtfulness that decenter the agency of the human subject.[6] This doubtful subject is not recognizable as a unitary individual or even as exclusively human but is a composite subject whose collaborations with algorithms open onto an undecidable future, where one is permitted to ask new questions on the political landscape.

In contrast with a critique of algorithms as opaque and unaccountable, then, my cloud ethics poses the problem differently, tracing the necessarily cloudy and partial accounts that algorithms give as they generate themselves as agents in the world. When Haraway describes the partiality of the "modest witness," she is concerned with "telling the truth, giving testimony, guaranteeing important things."[7] The truthfulness of the account is thus grounded not in the secure authorship of the reasoning subject but precisely in the impossibility of self-knowledge that anchors the account. I return again to Judith Butler's important warning on the partiality of all forms of account. "My account of myself is partial, haunted by that for which I have no definitive story," she writes, so that "a certain opacity persists and I cannot make myself fully accountable to you," this partiality and opacity being the condition of ethical relations.[8] In short, algorithms do not bring new problems of black-boxed opacity and partiality, but they illuminate the already present problem of locating a clear-sighted account in a knowable human subject. Here we begin to find the possibilities of an alternative orientation to doubt, one in which the subject necessarily doubts the grounds of their claims to know.

One place for my cloud ethics to begin to intervene, then, is in the giving of doubtful accounts, where the impossibility of a coherent account does not mark the limit point of ethics but instead describes the condition of ethico-political life. A doubtful account radically changes the terms of what counts as truthfulness in relation to the algorithm. If one relinquishes the search for a transparent truth of the algorithm, what kind of truth can be told? What kind of relation to oneself and to others is entailed by the algorithm's particular claims to the truth?

Ground Truths

As algorithms become increasingly pervasive in supplying solutions for state decision making, from the risk management of borders and immigration to decisions on outcomes in the criminal justice system, they hold out the promise of a particular claim to truth. In public forums, when algorithmic decision making is held to account, the form of truth is commonly professed to be an efficient and objective outputting of an optimized decision.[9] In fact, though, the mode of truth telling of contemporary algorithms pertains to the "ground truth": a labeled or an unlabeled set of training data from which the algorithm generates its model of the world. In a process of supervised machine learning, the algorithms learn from a ground truth model of data labeled by human domain experts, often via crowdsourced labor, such as Amazon's Mechanical Turk. As I describe in relation to the AlexNet image recognition algorithm in chapter 2, when a new set of features is extracted from the input data, these features are weighted in relation to the ground truth data. For example, a facial recognition algorithm used in urban policing is able to identify a face because of its exposure to a *ground truth* dataset of labeled images.[10] Increasingly, as deep learning algorithms derive their own ground truths by clustering raw unlabeled data, a model of what is normal or anomalous in the data is generated by the algorithms. The claim to truth made by machine learning algorithms, then, is not one that can be opposed to error or falsity, nor even to some notion of feigning that targets "fake news." Rather, the algorithm learns from the degrees of probabilistic proximity to, or distance from, a ground truth, itself often generated by algorithms. So, when neural network algorithms at the border or in policing or immigration reach a decision, this is a decision derived from an output signal that is entirely contingent on a set of gradients and weighted probabilities.[11] The architectures of neural network algorithms can contain twenty hidden layers, hundreds of millions of weights, and billions of potential connections between neurons.[12] Given the opacity of this arrangement, for an algorithm to generate an output (which is to reach a decision), it requires a mode of truth located in its relations to ground truth data. Put simply, the truth telling of the algorithm is entirely contingent on the notion of a particular kind of grounds that I call a *data ground truth*.

This malleable relation to a ground truth supplies the algorithm with the capacity to work with a grain of doubt and uncertainty. The output of the algorithm places a decision beyond doubt in the sense that it always already embodies the truth telling of the ground truth data. As one computer scientist explained of an image recognition neural network trained to detect cats, "it is not recognizing a cat, but a cat at this angle, that angle, in a tree, or upside down. . . . There is no sense of catness, just enough cats in places that there can be a pattern of what a cat is."[13] Thus, the decision that an entity is or is not a cat is placed beyond doubt, not in the sense that it could not be false but because it embodies the truth telling of the ground truth data of what a cat can be. Taking this truth logic to another domain of neural networks, one could similarly contest the output of a recidivism algorithm for being "false" in the sense that it wrongly assigns someone a high probability of reoffending, for example, but its degree of truth will always remain intact in its relations with the ground truth data. To be clear, the algorithm does not eradicate doubt, but neither does it only productively incorporate doubt, as has been observed in the methods of scenarios, catastrophe modeling, and preemption in geopolitics.[14] Though this is a science that can hold together multiple possible versions of events simultaneously, each possibility weighted as a layer of computation in the algorithm, it cannot live with doubt as such. How could a person say, "That is false; I am not in fact likely to reoffend if I am released," if the truth telling of the algorithm is anchored in the ground truth data and not in their situated life? That is to say, the machine learning algorithm must reduce the vast multiplicity of possible pathways to a single output. At the instant of the actualization of an output signal, the multiplicity of potentials is rendered as one, and the moment of decision is placed beyond doubt.

In the pages that follow, I address the practice of scientific truth telling in today's machine learning algorithms, and I seek to give some revitalized life to the embodied doubt that is always already present within the science. In one sense this is a form of critique of vision-dominated objectivity that claims to have a truth beyond doubt, but more precisely, it seeks something akin to what Donna Haraway has called the "embodied objectivity" of "partial perspective," and to what N. Katherine Hayles captures as an "embodied actuality."[15] This different orientation to doubt begins from the embodied doubts inhering within all notions of subjectivity and objectivity, doubting oneself and one's capacity to know. With this orientation to an embodied doubtfulness, a cloud ethics reopens the contingencies of the appearance of ground truth data, giving life to the fallibilities of what the algorithm has learned about the world across its billions of parameters, and rendering the output incomplete. Running against the grain of the Cartesian doubt of the fallibility of data derived from the senses, doubt in this alternative register is felt, lived, and sensed as embodied actuality in the process of an algorithm learning through its relations with the world.

Doubtful Science

To consider the doubtfulness of partial perspectives is not the same thing as casting doubt on the algorithm as such. Indeed, the point is that one could never stand outside the algorithm to critique or to adjudicate on its veracity since one is always already implicated in the algorithm as a form of adjudication of the truth. Instead, a cloud ethics dwells within and alongside the plenitude of teeming doubts in the experimental algorithmic model. To explain what might be at stake in this reorientation to doubt, I turn to an example drawn from a set of interviews with designers of deep learning algorithms that are actively generating their own ground truths about the world. The designers are explaining how the attributes within a ground truth dataset become the means to detect future attributes in other people, related only through the calibration of their attributes:

> One piece of work we did recently with an insurance company, where we were looking at the value of their life insurance customers. Now obviously they only take out life insurance maybe once, [laughs] so you haven't got lots of different data points that you can look at, but one of the things we can do there is (a), we can start to say, *well what does everybody else do?* And so we have what's called the context, so we're able to then compare the individual against a cohort group of attributes in order to be able to start to *make predictions.* And the other thing we can do (b) is we can take supplementary data in as well, so it could be, you know, a Facebook feed or an actuarial feed or it could be, you know, data that's coming through from the Home Office for instance, so we can start to supplement the data that already exists with other datasets in order to be able to improve our predictions.[16]

The value of a life insurance customer (the risk-based price of their insurance) is thus generated from the output of the neural network, calibrated against a threshold of value to optimize the insurance company's position. Where the data points on past known entities are scarce—as with the "courier" proxies for terrorists I discuss in chapter 4—the algorithm perceives what others are doing and establishes a ground truth from which attributes of risk or opportunity can be derived.[17] The life insurance algorithm emerges through the partial and oblique correlations among otherwise unrelated selves and others. It will go out into the world, continue to iterate and learn, ingest new data points, create new clusters, and generate new outputs for decisions. When the algorithm designers refine and adapt their algorithm for new domains, they are experimenting with the proximity between a specified target value and the actual output signals from their model, observing how the numeric scores generated by their model diverge or converge on the target. The computations of algorithms like these are doubtful in their openness to new interactions and encounters in the world, every layer infinitely malleable and contingent on plural interactions of humans and algorithms. A minor change in the statistical state of an entity—the gradient of a high-value life, or the features of a high-risk migrant—will transform the output signal and, thus, the targeted decision.

Understood as a practice that is partial, iterative, and experimental, the neural network algorithm is doubtful not only in the sense that it supplies a contingent probability for an absolute decision, but moreover because it actively generates thresholds of normality and abnormality. Put simply, the algorithm does not need to eradicate doubt or establish certainty for the decision because it generates the parameters against which uncertainty will be adjudicated. The algorithm learns to recognize and redefine what is normal and anomalous at each parse of the data. What does everybody else do? What are the attributes of the cohort? What is the value of the life? The response to such questions is never authored by a clearly identifiable human, but rather derived from a composite of algorithm designers, frontline security officers, experimental models of the mathematical and physical sciences, a training dataset, and the generative capacities of machine learning classifiers working on entities and events. For example, when Palantir supplies its neural network "investigative case management" algorithms to US Immigration and Customs Enforcement (ICE), the border guard ebbs into a composite of machine methods for "discerning a target" and "creating and administering a case against them."[18] The actions that are outputs of the algorithm—to arrest persons, to seize assets, to indict persons—are arranged in relation to the FALCON machine learning analytics engine and the ICE data warehouse.

There is a great deal at stake politically in the erasure of the doubtfulness of algorithms at the point of political decision. Though the making of a neural net algorithm for immigration enforcement or for life insurance risk is a fraught, political, and doubtful practice, the computational structure of such algorithms dictates that the final output must be a single numerical value between 0 and 1. In effect, the output of a neural network is a numeric probability, a single value distilled from a teeming multiplicity of potential pathways, values, weightings, and thresholds. This process of condensation and reduction to one from many allows algorithmic decision systems to retain doubt within computation and yet to place the decision beyond doubt. What would it mean to be able to express a posthuman doubtfulness in this context? How does one speak against the grain of the single output of automated visions of security? Is it possible to locate and amplify the doubtfulness dwelling within the partial fragments of the science itself?

Partial Accounts and Fatal Decision

To detail how partial accounts might advance a different orientation to doubt, I turn here to a historical moment when the output of a risk algorithm failed catastrophically. This is a moment when a Cartesian doubt was expressed that inaccurate data may have led to a catastrophic decision. When the event is understood as composed of opaque and partial fragments, however, what comes to matter is not whether something could be calculated accurately from data, but rather how partial probabilities became a singular calculus for a fatal decision. On January 28, 1986, seventy-three seconds after its launch at 11:25 AM, the NASA space shuttle *Challenger* broke apart over the Atlantic Ocean, killing all seven crew members, including Christa McAuliffe, NASA's "teacher in space." By February, President Ronald Reagan had established the Rogers Commission to "review the circumstances surrounding the accident and establish the probable cause or causes" and to "develop recommendations for corrective or other action."[19]

Among the fourteen commissioners appointed to the *Challenger* investigation was physicist Richard Feynman, who was persuaded to participate despite, by his own account, "having a principle of going nowhere near government" after his experience of working on the atomic program at Los Alamos.[20] More specifically, he was wary of bureaucratic reason and the governing of what he thought of as an unruly science by bureaucratic rules and protocols. The Cold War cybernetic rise of "algorithmic rules that could be executed by any computer, human or otherwise" with "no authority to deviate from them" had extended a particular kind of "arithmetic into the realm of decision," and this entanglement of mathematics and decision did not please Feynman.[21] Feynman's letters and diaries reveal a disdain for algorithmic decision procedures and axiomatic formulas, and a propensity to ask questions that ran against the grain of mathematical rules. "Doing it by algebra was a set of rules," writes Feynman of his early encounters with mathematics, "which, if you followed them blindly, could produce the answer."[22] For Feynman, algebra was a means of imposing an axiomatic set of procedures of rules on a puzzle that could otherwise intuitively work toward an unknown outcome.

Confronted with the Rogers Commission's setting of procedural steps in the investigation of the *Challenger* disaster, Feynman worked to reinstate the doubtfulness at each link in NASA's chain of events. He traveled to meet with the engineers, avionics scientists, and physicists whose data on particular components had made up the aggregate risk calculation on which NASA had based their launch decision. What he discovered was not a catastrophic departure from the normal rules or a "human error" or failure as such. Instead, the launch decision belonged properly to a posthuman composite of algorithms, where the steps of a normalized risk-calculation protocol had been followed beyond the limits of the calculable. As sociologist Diane Vaughan proposes in her account of the *Challenger* disaster, "harmful actions" can be "banal," and they can "occur when people follow all the rules."[23] Though Vaughan's famous account emphasizes the role of the situated culture of NASA in the errors made, Feynman's account foregrounds the doubt already present within the science of the program. Where NASA failed, according to Feynman, was in its tendency to "aggregate out" the multiple small fractures and failings that existed at the level of the most mundane of components and instruments. Understood in this way, Feynman's critical scientific method does not seek to correct inaccuracies but instead brings to the surface the doubts already present within each fragmented element so as to open up the breaches in the arrangement of propositions.

Here, I reflect on two aspects of Feynman's method for the possibilities they may offer to a critical cloud ethics method of partial accounts. The first is a particular *reinstatement of doubt* in data as they are given. The method of reinstatement is intended specifically to *reinstate*, or to give something back a position it had lost. This is of some significance, for doubt does not function to cast uncertainty on data that were heretofore settled and certain. Feynman's approach to doubt begins from the position that all scientific data are contingent, uncertain, and full of doubt. Indeed, among Feynman's major contributions to quantum physics was his "sum over histories" method, in which the calculation of particle interactions must "take into account a weighted tally of every possible chronology."[24] In this sense, Feynman's physics theorized precisely how a calculation might keep open the possibilities of multiple pathways that are otherwise incalculable. The algorithm unfolds as an array of mutable contingencies, each intimately and recursively layered with others so that the science strays from calculable paths and experiments with the incalculable. As Karen Barad suggests in her account of the intimacies of feminist science studies, "Life is not an unfolding algorithm" so that "electrons, molecules, brittlestars, jellyfish, coral reefs, dogs, rocks, icebergs, plants, asteroids, snowflakes, and bees stray from all calculable paths."[25] In my refiguring of doubt, I envisage the unfolding algorithm as straying from calculable paths in the manner Barad proposes. The hesitant and nonlinear temporality of the etymology of doubt, from the Latin *dubitare*, suggests precisely a hesitation, an uncertainty, and a straying from calculable paths. To be doubtful could mean to be full of doubt, in the sense of a fullness and a plenitude of other possible incalculable paths.

What might a method of reinstating doubt, or giving doubt a presence that it had lost, look like? Returning to Feynman's investigations into the *Challenger* disaster, the seals of the solid fuel rocket boosters of the shuttle were secured by pairs of rubber O-rings. It was known to NASA that, during some launches, the hot gases from the boosters could leak from the seals, causing what was called *blowby* over the liquid oxygen and nitrogen of the orbiter's engines. Though the engineers observed corrosion of the O-rings over time, they assumed that the heating of the rubber O-rings during launch caused small expansions to close the gaps in the seal. At the fatal launch, the temperature was 28 degrees Fahrenheit (-2 Celsius); the coldest launch recorded prior to this had been 53 degrees Fahrenheit. On the evening of January 27, the night before the launch, the engineers at Thiokol, which manufactured the seals, warned NASA that the launch should not take place if the temperature was below 53 degrees. Yet, the public record shows that NASA "took a management decision" because "the evidence was incomplete" that "blowby and erosion had been documented above 53 degrees," so "temperature data should be excluded in the decision."[26] As Feynman writes, NASA testified that "the analysis of existing data indicated that it [was] safe to continue flying."[27] In effect, the aggregate data of twenty-four past flights without a mission failure had placed the launch decision beyond the doubt that would otherwise have been reinstated by attentiveness to the components, and to the inscribed traces they embodied. A sense of doubtfulness that might be considered fully posthuman—dwelling in the material marks of blowby and erosion, and in the touch of engineers on fissures and cracks—was aggregated out in the calculation of a singular output.

In historical instances such as Feynman's account of *Challenger*, we can see some of the fallacies of an objective risk-based or data-driven decision placed beyond doubt. Remarking on the computer model used by NASA in support of their launch decision, Feynman writes, "It was a computer model with various assumptions. You know the danger of computers, it's called GIGO: garbage in, garbage out."[28] To question the assumptions of the computer model, Feynman began to focus on the fragments of data and their associated probabilities, reinstating doubt within each element. Understood in this way, the probabilistic calculation of risk gives way to what Donna Haraway has called "partial, locatable knowledges," in which there is the "possibility of new webs of connections."[29] NASA had testified to the commission that the probability of failure of a mission was calculated to be 1 in 100,000. "Did you say 1 in 100,000?," queried Feynman. "That means you could fly the shuttle every day for 300 years between accidents?"[30] To open up a breach in the risk calculus, Feynman gathered the scientists who had worked on the various different components of *Challenger*, asking them to estimate the probability of failure of the shuttle and to write this probability on a slip of paper. When he collected the fragments of paper with their pieces of data, he found that the calculus reflected the particular and situated relationship to a material component and its properties. Thus, for the seal engineers, the probability of failure was felt to be 1 in 25; for the orbiter's engines, 1 in 200; with none of the data elements showing NASA's aggregate probability of 1 in 100,000. Feynman's point was not to correct the calculation but rather to dramatize its impossibilities and incalculables. From the paper fragments of each embodied likelihood, gathered as incongruous scraps, the possibility of new webs of connections emerges.

The second aspect of Feynman's method I discuss here is the *affordance of capacities* to technical devices—in particular, an affordance of the capacity to give accounts of themselves, their breaches, and their limits. One approach to such affordances would be to say that, put simply, when something fails, it also speaks to us and tells us something of its limit points.[31] Yet, this is not merely a question of how the technical object has capacity to "gather to itself" a community of other human and nonhuman beings.[32] Instead, as Katherine Hayles reminds us, "Cognition is much broader than human thinking," so that cognitive capacities are afforded to other life forms, animals, and technical devices, forming a "rich ecology of collaborating, reinforcing, contesting and conflicting interpretations."[33] Within this broad ecology of cognition, the always present ethicopolitical difficulty of humans giving an account of themselves extends also to the partial and incomplete accounts of other beings.

When Feynman found that the Rogers Commission was unwilling or unable to register his critical account, he brought with him to the public assembly part of his material scientific community, inviting the material device of the O-ring to give an account in the forum.[34] As the government officials and press gathered for a meeting with the commission, Feynman placed a piece of rubber O-ring in a clamp bought in a downtown hardware store. Requesting a glass of ice water, he dropped in the materials and waited for them to cool. When the rubber was removed from the glass, brittle and compressed, it broke apart before the assembled audience (figure 5.1). One reading of the event of

Figure 5.1　Richard Feynman removes the O-ring material from ice water at the Rogers Commission inquiry, February 11, 1986. "Richard Feynman (Articulate Scientist)—Space Shuttle Challenger Testimony," https://www.youtube.com /watch?v=MWZs8l2AMps, posted October 3, 2013.

the O-ring giving an account of itself is that Feynman follows the "structure of heroic action" in science, as Donna Haraway has put it, speaking objectively and "self-invisibly" through the "clarity and purity of objects."[35] Certainly the now famous *New York Times* report of the O-ring event, citing Feynman's few words, "There is no resilience in this material when it is at a temperature of 32 degrees," bears a form of witness close to that of Haraway's critique. I locate a different form of account here, however, one with no unified authorial source or transparent picture of the truth, but rather a distributed and oblique account in which we see the political impossibility of resolving the truth before the public. What is dramatized with the rubber O-ring in the ice water is a set of relations that are material and political. Refusing the mode of accountability in favor of this different form of giving an account, there is what Karen Barad calls a "condensation of responsibility" in matter, wherein the multiple contingent past decisions are lodged within the object, engaging us in "a felt sense of causality."[36] In this way, a material limit that was ordinary in the embodied experiences of engineers and materials scientists—familiar to their touch—could be expressed as a claim that could be heard in the world. Feynman's claim did not express the voice of Haraway's "victor with the strongest story"; indeed his account was derided and rejected by the commission throughout.[37] At the specific moment I describe, Feynman speaks against the grain of Cartesian doubt and instantiates a mode of doubtfulness more commonly feminized and annexed within subjectivity. It is precisely this mode of intuitive causality and embodied doubtfulness that I am seeking as a resistant and critical form of responsibility.

Risky Speech

To speak of the excesses and limits of technoscience in our contemporary present has become extraordinarily difficult. Every output of the algorithm, even when it leads to wrongful detention or racialized false positives, is productively reincorporated into the adjustment of the weights of a future model. The method of reinstating doubtfulness within algorithmic arrangements, however, has the potential to yield a kind of "risky speech," or "parrhesia," where, as Michel Foucault contends, an account is given that places itself at risk.[38] Significantly, Foucault considers it possible for science to give parrhesiatic accounts, suggesting that "when deployed as criticism of prejudices" and "current ways of doing things," science can place itself at risk and play the "parrhesiatic role."[39] For Foucault, the defining character of parrhesia is that the parrhesiast is bound to the truth of their claim, but they place themselves at risk in so doing, because they speak against the grain of prevailing thought. In this sense, the parrhesiast can never "witness with authority" but only with a curiosity that is not counted before the public assembly. As Judith Butler comments on Foucault's mode of criticism, the practice "involves putting oneself at risk," "imperilling the very possibility of being recognized by others," and risking unrecognizability as a subject."[40] Understood thus, risky speech does not come from an already authorized subject, but only from a subject who risks making themselves unrecognizable to others. To place oneself at risk of being unrecognizable is what seems to be necessary in an age of algorithmic calculations. What marks out the risky speech of parrhesia from other forms of political claims, other modes of discursive and performative speech, is its capacity to open onto an indeterminate future: "There is a major and crucial difference. In a performative utterance, the given elements of the situation are such that when the utterance is made, the effect which follows is known and ordered in advance, it is codified. . . . In parrhesia, on the other hand, the irruption determines an open situation, or rather opens the situation and makes possible effects which are, precisely[,] not known. *Parrhesia* does not produce a codified effect; it opens up an unspecified risk."[41]

Understood in these terms, parrhesiatic speech is an irruption that opens a situation and makes possible effects that are not known in advance. By contrast, the search for an encoded ethics of algorithms proceeds from a known subject and a problem that is ordered in advance. For example, when Google employees protested Project Maven's "involvement in war" and "biased and weaponized AI," the problem of "data captured by drones to detect vehicles and other objects" was ordered in advance. Despite the manifest entanglements of the drone program in multiple other Google object recognition systems, the protest delineated good from evil, and war from commerce, in ways that did not place the subjects and objects of science at risk. Within a cloud ethics, risky speech permits loosening the hold on subjectivity and relinquishing the grounds from which the subject speaks. When the "avowing subject," as Foucault writes, "loosens its hold," it becomes possible to envisage a risky speech that forges new ethicopolitical relations with oneself and others.[42] This means that one could be doubtful of all claims, for example, that the bias or the violence could be excised from the algorithm and begin instead from the intractable political relations between the algorithm and the data from which it learns.

Ultimately, on discovering that the Rogers Commission report contained a previously undisclosed "tenth recommendation," sheltering the risk algorithm within "error" and commending NASA's work and mission to the nation, Richard Feynman refused to add his signature to the report. His fellow commissioners sought to persuade him, suggesting that the tenth recommendation was a feminized "only motherhood and apple pie," and that "we must say something for the president."[43] The tenth recommendation, which would authorize a particular decided future of continued NASA missions, was demarcated from the real science of objective findings and rendered maternal and nation building.

In response, Feynman asked what kind of truth could be spoken to the government: "Why can I not tell the truth about my science to the President?" In the event, his own report appeared as an appendix to the published document, titled "personal observations on the reliability of the shuttle," as though his merely personal situated account could only be subjective and without grounds. Feynman's appendix is risky speech in Foucault's sense, an argument that places the subject at risk, runs against the grain of the commission report, and challenges the form of truth telling of the public report. In this sense, to be doubt*ful* is to open onto the full contingency of a situation and to be responsive to unspecified future effects. The fullness of an embodied posthuman doubt I envisage here is a critical cousin to Lauren Berlant's notion of a "cruel optimism," where a "cluster of promises could seem embedded in a person, a thing, an institution, a text, a norm, a bunch of cells, smells, a good idea." The cruelty Berlant so vividly describes is a "relation of attachment to compromised conditions of possibility whose realization is discovered to be impossible, sheer fantasy, or too possible, and toxic."[44] Though Berlant does not probe directly what it might mean politically to be doubtful of the promises as they embed in a person, a thing, or a technology, her work does envisage the making of claims in the present so that there remains "the possibility of the event taking shape otherwise."[45] Understood thus, to be doubtful could be to experience a fullness or multiplicity of the present moment and the many ways it might unfold, such that the cruelly optimistic promises of technoscience do not cling so tightly to ideas of the optimal or the good enough output. The optimism of the algorithm is founded on its optimization, that is, on its capacity to reduce the fullness of the present moment to an output to be acted on in the future.

Doubtfulness and Ethics

If a signaling of the multiple errors and faults in contemporary machine learning algorithms cannot meaningfully provide grounds for critique, then can there be risky speech amid algorithmic techniques? I wish to note here that throughout all my discussions with algorithm designers, there has been a curious mood of twinned optimism and doubt: optimism that a good enough model can always be found, and doubt that the "junk data" elements of social media sentiment or unlabeled image files could ever be adequately prepared.[46] If, however, the many misidentified faces in biometric systems, or wrongly seized assets in a Palantir immigration system, merely "unleash computable probabilities into everyday culture," as Luciana Parisi reminds us, then how does one speak against the grain of prevailing algorithmic logics?[47] Put differently, Richard Feynman excavated and amplified the doubt within the analog data elements by painstakingly reconstructing the plural probabilities of each finite piece. If this method of gathering calculative fragments allows us to pause and dwell within the doubtfulness of the calculus, can we imagine today an equivalent method of asking the scientists we research to please "write the probability of the failure of your piece of the machine learning software on this piece of paper." How would the reinstatement of doubt account for the adjustments of weights that are conducted by the algorithm on itself? If one sought to reinstate the doubtfulness of test data, then what kind of truth-telling practice could this be? How would it extend to the Amazon Turk out-sourced workers who had labeled the images for a machine learning algorithm to recognize a future human face? Could the doubtfulness of a numeric prob-ability account for the future uses of one algorithm used to teach another?

When Richard Feynman asked the scientists on the shuttle program why the fragments of their doubts had not found their way into the future calcula-tions of risk assembled by NASA, he was informed that "it is better if they don't hear the truth, then they cannot be in a position to lie to Congress."[48] The problem of truth telling in this instance lies less with science as such, and more with the various ways in which embodied doubtfulness in science does not give political accounts of itself. Similarly, when the former US director of national intelligence James Clapper was found to have lied to Congress on the analysis of data on US citizens in March 2013, he later argued that he had provided the "least untruthful answer possible in a public hearing."[49] To speak against the grain of the promise algorithms make in the name of political resolution is to bring to the public assembly a kind of contemporary heresy. Following the UK terror attacks in Manchester and London in 2017, for example, where some of the perpetrators had previously been known to intelligence agencies, the offi-cial inquiry by David Anderson has concluded that what is required is a more effective deep machine learning algorithm for the identification of high-risk individuals among twenty thousand so-called closed subjects of interest.[50] In effect, this implies training an algorithm to recognize the attributes of known individuals in the dataset to identify similar attributes among unknown fu-ture subjects. The weighted probabilities among the twenty thousand will be used to generate target outputs for future security action. To signal the doubt-fulness of such an algorithm is precisely to engage in risky speech and annul the promise of a securable future.

The form of doubtfulness I have described would demand that algorithms give political accounts that are marked by contingency and ungroundedness. In place of an ethics that seeks defined grounds for algorithmic decisions, a cloud ethics begins from the ungroundedness of all forms of decision, all po-litical claims, human and algorithmic. As Thomas Keenan writes, "What we call ethics and politics only come into being or have any force and meaning thanks to their ungroundedness," so that "we have politics" precisely because we have no "reliable standpoints." If a future political claim is to be possible, then "its difficulty and its persistence stem from its terrifying, challenging, re-moval of guarantees."[51] If there are to be future claims to human rights not yet registered or heard, for example, then there must always also be doubts haunt-ing the possibility of rights, doubts that acknowledge the irresolvability of all

Figure 5.2 The threat centaur model of intelligence gathering and threat detection, as presented by Recorded Future. Author photograph.

political claims. Here, doubt pervades what Keenan describes as a "darkened frontier," where the essential difficulties of politics must take place, and where all decisions must, following Derrida, "act in the night of non-knowledge and non-rule."[52] In holding out the promise of resolving political difficulty—coming out of the darkness with the target output of a machine learning algorithm—technoscience harms the terrain of the political. The algorithm promises to resolve the political by holding together a multiplicity of possibilities within the neural net, while carving out a space for a single optimized decision. This is not to say that the decision is made without doubt. On the contrary, it is doubtful, full of doubt, and yet it decides with indifference.

Seated in the audience at a UK technology company's presentation, I watched the pitch made to UK government officials in the name of resolving political difficulty with machine learning algorithms.[53] Recorded Future explained how their "threat centaur" machine learning algorithms (figure 5.2) would "scrape the web for geopolitical events" and send "real time alerts" on "political protests, terrorist attacks, and civil unrest."[54] As the algorithm designers describe the system, it deploys a combination of rule-based and deep learning algorithms. The rule-based algorithms are said to be based on "human intuition" about whether "an entity is associated with some kind of risk," while the machine learning algorithms are trained on a "large dataset, using trusted threat list sources as ground truth for what constitutes a malicious entity." The figure of the "centaur" in the algorithmic system is a posthuman body with a capacity to reach an unsupervised algorithmic decision ("this event is critical") but also to display "a human readable motivation for that judgement," so that action can be authorized against a threat.

At one level the algorithms of the threat detection centaurs appear to run counter to the form of risk calculation observed by Feynman at NASA. The smallest of infractions or the subtlest of signals of a component failure or a change in online clickstream patterns, for example, is said to yield a threat alert to the centaurs. Like the Rolls Royce jet engines that yield real-time data on the performance of aircraft as they fly, the twenty-first-century O-ring would almost certainly be given a data voice. And yet, at the level of practices of truth telling, there is something of significance here. The *Challenger* launch decision teemed with a cacophony of embodied doubt—on the part of the engineers, physicists, and mathematical modelers, in communion with their nonhuman partners—and yet, the decision was placed beyond doubt. What we are seeing emerge in contemporary algorithmic technoscience is an orientation to truth telling that says that all political difficulty can be resolved, all decisions can be placed beyond doubt.

How might one intervene in models such as the threat centaur in ways that reinstate the doubtfulness of the algorithm? It is necessary to engage the traces of machine learning algorithms, to be curious about them, to doubt them, but also to listen to the doubt they themselves express in the world. In engaging research methods that follow elements of an algorithmic solution as they travel and have onward life, I am trying to be a receptive listener when the piece of data appears to say "you do know that I cannot really be cleaned; I remain in there, muddying the calculation"; or when the biometric-matching algorithm says "in 14 percent of cases, where ambient lighting exceeds these parameters, I cannot be read"; or when the pattern recognition algorithm says "the training dataset I have been exposed to has established what looks normal or abnormal to me." There is doubt, and it proliferates everywhere; at every step in the application of algorithmic solutions to political questions, doubt lives and thrives and multiplies. And, where there is an ineradicable doubt, there is also the darkened frontier and, therefore, the potential for politics. Inspired by Richard Feynman's method of reinstating doubt in the data points, and being an interlocutor for the people and things "in the engine" who actively give accounts of limit points of all kinds, I urge that we experiment as the algorithms do, iteratively and recursively, giving *doubtful accounts* of the calculation.

The Open Channel

To speak in praise of an embodied form of posthuman doubt is to draw together several things that do not sit together easily in terms of critique. The form of doubt I am proposing is to be distinguished from Cartesian skepticism, for it does not seek to annex fallible sensory doubts in the pursuit of foundational truths. As posthuman composite life forms, the doubtful subjects I depict here dwell uneasily within partial and situated knowledges, and when they make a claim in the world, they do so in ways that stray from calculable paths. I include within this category of doubtful subjects all those whose science bears a responsibility for an apparently risk-free, error-tolerant political decision, from drone strikes to border controls, and from voter targeting to immigration decisions. In so doing, I follow Foucault's sense that the scientist and the geometer do not stand outside the capacity for parrhesiatic, or risky, speech, alongside Karen Barad's caution that we are materially immersed in and inseparable from science. Doubtfulness expresses the many ways in which algorithms dwell within us, just as we too dwell as data within their layers, so that we could not stand apart from this science even if we wanted to. The doubtfulness of our relations to ourselves and to algorithms implies that critique will always involve "putting oneself at risk," as Judith Butler describes it, and risking "unrecognizability as a subject," as one can "never fully recuperate the conditions of . . . emergence."[55] In this way, the unrecognizability of the subject extends into the opacity of the algorithm, so that new forms of subjectivity emerge that are never fully recuperated. The next time one hears that the "black box" of the algorithm should be opened, one might usefully reflect on the unrecognizability of all forms of self.

To express doubt in these terms is to try to find a means to respond to algorithms that incorporate doubt into computation, converting doubts into weighted probabilities that will yield a singular output that can be actionable. A critical response cannot merely doubt the algorithm, pointing to its black-boxed errors and contingencies, for it is precisely through these variabilities that the algorithm learns what to do. Instead, an orientation to doubtfulness decenters the authoring subject, even and especially when this author is a machine, so that the grounds—and the ground truths—are called into question. Understood in this way, to doubt is also to reopen a "decision worthy of the name," as Derrida notes, or to open onto "a politics of difficulty," as Thomas Keenan observes, in which the "break with the humanist paradigm" does not close off the ethicopolitical but provides its starting point.[56]

What could it mean to value doubtfulness as a critical faculty, to reinstate doubt into the composite creature that is automated algorithmic calculation? It is not the same thing at all as calling for an ethics of human decision in machinic security (putting the human back in the loop), or saying that algorithmic decisions are intrinsically malevolent. Instead, it expands the space for doubt beyond what Deleuze, following Bergson, calls "the decisive turn," where lines diverge according to their differences in kind.[57] Understood in this way, the action signaled by the output of the algorithm is never placed beyond the darkness of doubt and difficulties, for it carries doubt within, is always incomplete, and does not know what is around the decisive turn. Though contemporary algorithms reduce the multiplicity to one at the point of decision, they do not leave behind the knots of doubt that dwelled within the multiple. The condition of possibility of a condensed algorithmic output, in short, is a teeming multiplicity of potentials that remain in play. As Barad writes on Feynman's quantum field theory, in which the electron absorbs within itself its own proton, and "there is something immoral about that," the action of the body is never fully beyond doubt.[58]

In so many ways, of course, the physicist Richard Feynman is a curious figure to invite into conversation with Haraway, Braidotti, Hayles, and Barad on science and the ethicopolitical, particularly given his role in the material algorithmic violence of the nuclear age. Yet, the point is that none of us can stand outside this science and judge it to be good or evil, to say definitively "things should not be this way." We are within it, and it is within us, such that critique cannot begin from an outside to the algorithm. As Barad suggests, "Ethicality entails noncoincidence with oneself," so that to give an ethical account is to resist anchoring the claim in a knowing subject.[59] Doubt does not stand outside and pronounce judgment; doubt is also interior, essential to the displacement of all subjectivity, for we doubt ourselves and our place in the world. For Feynman, doubt is what he describes as the "open channel" in politics and in science: "If we take everything into account . . . then I think we must frankly admit that *we do not know*. But, in admitting this, we have probably found the open channel. . . . The openness of possibilities was an opportunity, and doubt and discussion were essential. . . . We must leave the door to the unknown ajar. . . . It is our responsibility to leave the people of the future a free hand."[60] To leave the door ajar is to resist the promises made that the future is resolvable through the optimized output of algorithmic decision engines. I open this book with a discussion of the algorithm as a "hinge" on which the very idea of politics might turn. To leave the door to the unknown ajar is to reopen the many points in the algorithm where another future was possible, where the hinge does not quite limit the possible movement. While the enfolded doubts of the algorithm's feedback loop and back propagation deny the impossibility and difficult politics of decision, an embodied doubt pauses with the undecidability of alternative pathways and their contingent probabilities. The claim to a ground truth in data that pervades our contemporary geopolitical imagination precisely closes the door to the future, offering algorithmic solutions to resolve the difficulties of decision. To reinstate embodied doubt within the algorithm, and to allow the components of the badly formed composite to speak of their limits, is to seek to leave the door ajar for the making of future political claims, for these future unknown claims can never appear in the clusters, attributes, and thresholds of the algorithm.

The Unattributable
Strategies for a Cloud Ethics

I return to the question of apartheid. It is exemplary for the questions of responsibility and for the ethical-political stakes that underlie this discussion. . . . This is the moment of strategies, of rhetorics, of ethics, and of politics. Nothing *exists* outside context, as I have often said, but also the limit of the frame or the border of the context always entails a clause of nonclosure.
    —Jacques Derrida, *Limited Inc.*

A description must capture not only the objects contained in an image, but it must also express *how these objects relate to each other* as well as *their attributes* and the *activities* they are involved in.
    —Oriol Vinyals et al., "Show and Tell: A Neural Image Caption Generator," my emphasis

"Show, Attend and Tell"

At a major international machine learning conference in 2015, a group of Google computer scientists presented their model for addressing what was, at that time, one of the most difficult and important problems for deep learning: "scene understanding."[1] The algorithmic model they called "show, attend and tell" is one of several experiments in the use of generative neural networks for representing and interpreting what is taking place in a scene.[2] The machine learning model combines two sets of algorithms (convolutional neural networks, or CNNs, and recurrent neural networks, or RNNs) that learn together how to generate a description of the most likely thing of interest in a scene, its attributes, and its future actions. Giving an account of their "attention framework," the designers explain how the first set of algorithms—CNNs for extracting the feature vectors of a scene—"distill information in an image down to the most salient features."[3] In this process of distillation, the algorithms are learning how to "infer scene elements" in a crowded data array and to foreground what is considered to matter among the relations of people and objects.[4] This foregrounding is a mode of attentiveness that simultaneously deploys its contextual data environment to learn and discards that context as background noise. In this sense, it is a form of modern attention that necessarily also involves practices of distraction and annulment.[5] Once the first neural networks have extracted the feature vectors for what matters in the scene, a second set of algorithms—RNNs designed for natural language processing—then act on the output of the first to "decode those representations in natural language sentences."[6] Working together in collaboration, the algorithms are said to generate outputs deciding "not only the objects contained in an image," but also how they "relate to each other," what their essences or "attributes" are, and what probable "activities" they may be "involved in."[7] The training images of the algorithms—supplying the features of the recognizable—thus display a curious and compelling politics of showing and telling for our times.

Let us reflect on how these algorithms are condensing the question of what matters, "inferring the scene," as the computer scientists describe it, and "mimicking the remarkable human ability" to filter out "the clutter" and attend to the thing of interest.[8] In figure 6.1, a group of people gathers in a marketplace, their relations uncertain and indeterminate. One cannot possibly know from the image how these bodies relate one to another, what their qualities or attributes might be, and especially why they are there. Indeed, perhaps the incalculability and unknowability of their relations is the defining character of the urban scene, the crowd, and the marketplace. And yet, from the multiplicity of relations in the scene—the people, the vegetables, the earth, the clouded backdrop of market stalls—the neural nets cluster the data and define the attributes necessary to generate a single condensed output of meaning: "A group of people shopping at an outdoor market. There are many vegetables at the fruit stand." The algorithms generate the field of meaning; they are the mise-en-scène of the event; they decide *what is to be* in the single output signal. Understood in this way, there is indeed nothing outside the text in the neural net, the context supplying the feature space from which all possible future attention can be extracted, all meaning derived. Simultaneously, though, as the algorithms mistakenly infer the "vegetables" at the "fruit stand" from all the past data elements of images and video of markets they have been trained on, the "outside penetrates and determines the inside," so that the borders of context entail also an opening, or nonclosure.[9] As the al-

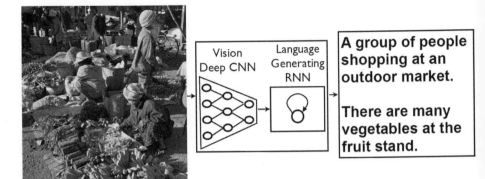

Figure 6.1 The *show and tell* model for scene understanding, comprising two algorithms: a convolutional neural network for image recognition and a recursive neural network for natural language generation. LeCun, Bengio, and Hinton, "Deep Learning," 2048.

gorithm travels from its laboratory site of test images of South Asian markets, how might the presence of a particular object in the scene—a turban, some plantain, a pile of baskets—imply the attributes for future instances in Middle Eastern marketplaces? If the Google computer scientists' show and tell algorithms are animating the scene understanding of the US drone program, what action is authorized based on attributes?[10] When algorithms learn the "latent alignments," what happens to that which is considered to be dormant or lying hidden within a scene? What is a normal alignment in the marketplace, the border landscape, or the city plaza?

The techniques deployed within the neural net algorithms to condense the features of a scene to a single output—"a woman is throwing a Frisbee in a park," "a stop sign is on a road with a mountain in the background"—give a compelling account of the ethicopolitics of algorithms for our times (figure 6.2). The output of the algorithms reduces the intractable difficulties and duress of living, the undecidability of what could be happening in a scene, into a single human-readable and actionable meaning. We have ethical and political relationships with other beings in the world because the meaning of those relations, their mediation through every scene of life, cannot be condensed. They are precisely irreducible. And so, at the moment that the algorithm outputs a single meaning from an irreducible scene, at this border limit is also a "clause of nonclosure," as Derrida describes the opening of context.[11] The ties that bind words, concepts, images, and things are only ever relatively stable, so that there is always also an opening. How does one begin to locate the points of nonclosure within the algorithm's program of meaning making? If a cloud ethics is to be concerned with the political formation of relations to oneself and to others, how might it locate alternative critical relations of showing and telling? Are there counter-methods of attention available to us that could resist the frameworks of attention of machine learning? Amid the technologies of the attribute, what remains of that which is unattributable in the scene?

In this concluding chapter, I am concerned with how a cloud ethics could be brought into being in a world that is increasingly shaped through the arrangements of the machine learning algorithm. Though I do not propose a definitive method for critique or for resistance to the prevailing powerful logics of surfacing, targeting, or "show and tell" in machine learning, I do urge

A woman is throwing a <u>frisbee</u> in a park.

A <u>dog</u> is standing on a hardwood floor.

A little <u>girl</u> sitting on a bed with a teddy bear.

A group of <u>people</u> sitting on a boat in the water.

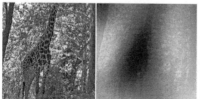
A giraffe standing in a forest with <u>trees</u> in the background.

A <u>stop</u> sign is on a road with a mountain in the background.

Figure 6.2 The *show, attend, and tell* model for scene understanding, displaying the test image on the left of each pair, with the machine's "attended" vision on the right. The target output is underlined. Xu et al., "Show, Attend and Tell," 2.

that it is possible to confront the specific fabulatory functions of today's algorithms with a kind of fabulation of our own. "It is the task of the fabulatory function," writes Deleuze, "to invent a people," to write into being a collective or a people who are otherwise "missing."[12] Now, of course, algorithms engage in fabulation—they invent a people, write them into being as a curious body of correlated attributes, grouped into clusters derived from data that are themselves fabulatory devices. Indeed, the fabulation of the algorithm involves a form of iterative writing, a collective enunciation of which we are all a composite part. To consider the *writing* of algorithms, as I argue in this book, is to locate the ethicopolitical acts of relating narratives of oneself and others. So, the methodology of a critical cloud ethics must also fabulate, must also necessarily invent a people whose concrete particularities are precisely unattributable, never fully enclosed within the attribute and its output. To fabulate in this way means to take the arrangements of the algorithm (in this instance, the show and tell neural nets) and imagine their mode of existence beyond the decisive point of the target output. As method, a cloud ethics engages fabulation in that it does not anchor narrative in the authorial source but rather "digs under stories, cracking them open," so that the reader might enter into the space of the becoming-political of the algorithm.[13] What this means is that one cannot stand outside the algorithm to judge its morality, its role in doing good or evil. Instead, one must begin from the iterative writing that is itself generative of fungible thresholds of the good and the bad.

Just as computer science has become comfortable with experimental fabulation and narrative, it is necessary for others to play with the arrangements, thresholds, and assumptions of algorithms. The computer scientists are, for example, apparently comfortably drawing on philosophy and ethics in their assertions that machine learning contains within it a capacity for something like machine reasoning.[14] Methodologically, we social scientists, and humanities and ethics scholars, must also find ways to engage experimentally in algorithmic fabulation. Let us begin by digging under the narratives of the show and tell algorithms, cracking them open to find the moments of nonclosure, writing them in ways that lever open their ethicopolitical orientations. Put simply, as algorithms work to derive future propensities from collections of attributes, so we too must curate the attributes of the algorithm to signal its propensities to colonize the unattributable future. A fabulatory method would take the arrangements, perceptions, and accounts of the algorithm and rework them to deepen the political intensity of what it means to output something into the world (figure 6.3):

A large white <u>bird</u> standing in a forest.

A woman holding a <u>clock</u> in her hand.

A person is standing on a beach with a <u>surfboard.</u>

A woman is sitting at a table with a large <u>pizza.</u>

A man wearing a hat and a hat on a <u>skateboard.</u>

A man is talking on his cell <u>phone</u> while another man watches.

Figure 6.3 The errors in the algorithm are used to gain intuition into what the model saw and how it perceived the scene. Xu et al., "Show, Attend and Tell," 5.

A woman is throwing a Frisbee in the park.
A woman is holding a child at the border fence.
A stop sign is on a road with a mountain in the background.
A placard is held in Tahrir Square with an unauthorized crowd in the foreground.
A man is talking on his cell phone while another man watches.
A man is opening his rucksack while another man watches.
A group of people shopping at an outdoor market. There are many vegetables at the fruit stand.
A group of people running at an outdoor market. There are many weapons in some hands.

To advance a cloud ethics is to work with the propensities that the algorithm embodies.[15] In place of an ethics that seeks to make an illegible algorithm legible to the world, a cloud ethics recognizes the nonclosure in all forms of writing and pushes the fabulation of the algorithm beyond what can currently be read of its logic. The single output of the algorithm is reopened and reimagined as an already ethicopolitical being in the world. What could this small adjustment in the weighting of the CNN do to the language of the output meaning? What is a "salient feature," and how does it come to the forefront of the crowded scene? In the following sections, I address the potential strategies of a cloud ethics across three aspects: *apertures*, *opacity*, and the *unattributable*.

Apertures

The computer science problem of how to understand the space of a scene—to show, to attend, and to tell what is happening—is not limited to the generation of meaning from images. Indeed, for machine learning algorithms, all space is feature space. That is to say, whether the scene is aerial video from a drone or images from an MRI scan, the algorithm acts on a feature space understood as the n-dimensions where data variables reside. The neural networks that extract the *salient features* from images, for example, do not meaningfully differentiate the sources of their input data. At the level of the algorithm, the image-derived input data that generate a given meaning (e.g., "This is a cat," or "A placard is held aloft in the city square") take the form of numeric values for the presence or absence of edges in the pattern of pixels.[16] It would make little technical difference if the input data to the algorithm included other forms of data, such as the sentiment analysis of social media text, travel data, or cell phone location data. The neural network converts all these data into gradients or feature vectors that can then be recognized as similar to, or different from, the feature vectors present in other known scenes or situations. What matters to the algorithm, and what the algorithm makes matter, is the capacity to generate an output from the feature vectors, to be able to "tell" what is latent in the scene. For example, when Facebook supplied their users' data to Cambridge Analytica for the generation of models of voter propensities, they supplied the feature vectors necessary to train algorithms to "tell" what is latent in the democratic political "scene."[17] It is precisely this emerging alliance— between algorithms hungry for new feature spaces to yield attributes, and a modern politics anticipating the latent attributes of populations—that poses one of the principal ethicopolitical challenges of our times. What is at stake is not only, or not primarily, the predictive power of algorithms to undermine the democratic process, or to determine the outcome of an election, a judicial process, a policing deployment, or an immigration decision. Of greater significance even than these manifest harms, algorithms are generating the bounded conditions of what a democracy, a border crossing, a protest march, a deportation, or an election can be in the world.

I do not wish to understate the potential that algorithms have for profound cruelty and violence. Indeed, when I say that the harms of algorithms are not principally located in the much-publicized invasions of privacy or absence of transparency, I am identifying a profound violence in the algorithm's foreclosure of alternative futures. This is a different kind of crisis of democratic politics, one in which deep neural networks do not merely undermine the already decided functions of democracy ("decide the US president," or "decide to leave the European Union") but actively place a limit on the potential for any other form of political claim to be made in the future. What new entity or alternative model of citizenship could come into being after the algorithm has optimized all outcomes? What new political claim, not yet registered as claimable, could ever be made if its attributes are known in advance? These are not only philosophical questions but also practical ones with a real purchase on the politics of our times. Given the machine learning articulations I have followed in the researching of this book, one would have to contemplate, for example, whether a set of algorithms designed for the US presidential election actually learned to recognize the attributes of "clusters" of anti-immigration and nationalist sentiment via the "ground truth" data of the attributes of UK citizens who voted to leave the European Union. In short, the "salient features" of groups within the UK EU referendum may have gone on to have specific and violent ethicopolitical relations to future families who find themselves separated at the US-Mexico border.

What can a society be in the future if an algorithm is optimizing an output that imagines the resolution of all future claims? The multiplicity of potential pathways within the neural net—millions of potential alternative connections and correlations—is reduced to one, condensed to a single output. Understood in this way, contemporary algorithms are simultaneously constructing the architecture of the problem space (What should the state do at the border? How ought a society decide creditworthiness? What should a police force attend to?) and generating the solution to the problem. "The problem always has the solution it deserves," writes Deleuze, so that the solution (output) is generated by "the conditions under which it is determined as a problem."[18] If one understands politics as expressing the impossibility of resolving all problems with a solution, then algorithms appear to circumvent the political, because they adjust the parameters of the problem to experiment abductively with the "good enough" solution. Alternative political arrangements become more difficult to imagine, in part because—reversing Deleuze's formulation—the solution always gets the problem it deserves; that is, the propositional arrangements of the algorithm (the model of the problem) are immanently adjustable in relation to the target output (solution). If, for example, a CNN "shows" that the entity "vehicle" extracted from a video feed has a 0.65 probability of being part of a military convoy, and then an RNN "tells" that this is an actionable threat, at the narrowed aperture of the output there is no place for a 0.35 probability of a civilian bus in the scene. How could there be an alternative relation, a nonclosure, when the output has so intimately and abductively generated the model?

A strategy for cloud ethics would dwell for some time with the aperture of the algorithm, the point where the vast multiplicity of parameters and hidden layers becomes reduced and condensed to the thing of interest.[19] To be clear, the aperture is not a matter only of vision or opticality but, as Jonathan Crary has detailed in his histories of perception, is specifically a means of dividing, selecting, and narrowing the focus of attention.[20] The aperture of perception matters to my cloud ethics because it is a point of simultaneous closure and nonclosure. It is a closure in the sense that there is no outside of the algorithm, so that the weights and the outputs interpenetrate and mediate one another. The aperture of the algorithm is an ethicopolitical closure of all the alternative possible values, assumptions, and connections that the hidden layers may contain. And yet, at the limit of the aperture is also an opening onto a moment of decision that is always already fully ethicopolitical. From its roots in the Latin *aperire*, or to open, the aperture is an opening or breach through which light can enter. Understood as a form of aperture, the algorithm simultaneously filters down and closes off as it opens a gap or a breach. What this means in terms of a strategy of cloud ethics is that one can envisage attending to the teeming multiplicity of all past moments that have yielded their probabilities to the model. A cloud ethics opens up the space of condensed output and makes the decision difficult and freighted with the traces of rejected alternatives. The aperture is the moment at which space is divided and the multiplicity is reduced to one, and yet it must also contain within it the traces of the opening onto lost pathways. A cloud ethics alters the relation to a future that says "here is a good enough solution" by revaluing the aperture as a moment of decision in the dark, a calculation of the incalculable. The strategy involves reopening the multiplicity of the algorithm, digging under the stories, and attending to the branching pathways that continue to run beneath the surface.

How does one begin to locate the nonclosure in the arrangements of the algorithms I have discussed in the chapters of this book? When I have given lectures on cloud ethics, a number of generous computer scientists have asked questions from the audience, urging me not to pursue the notion of revisiting the weightings within a machine learning algorithm. For them, the adjustment of weights within the hidden layers is an impenetrable process that retains its opacity even to those who undertake it. "You cannot make the weights political, Louise," they have cautioned me, "because they are not really a thing. We don't know how they work. We are just messing around with them." Yet, it is exactly this kind of opaque, messy, and embodied relation to the weightings of the algorithm that interests me. Moreover, the sense of nonknowledge—where the designers of algorithms "don't know how they work"—is fertile ground for insisting on a cloud ethics that has to decide in the darkness of not knowing. Put simply, my cloud ethics would make the adjustment of a weight in a neural net weightier, heavier in the sense of carrying the burden of difficulty and intransigence. Such a heaviness of the weighting of probabilities could shift the ethical focus from the spatial arrangement of combinatorial possibilities to a temporal and durational force of becoming. The weighting within one layer of an algorithm becomes not only a spatial weight of probability but also a temporal weight of duration. There is a kind of injunction here: *the algorithm must carry the weight of its weightings.* As Deleuze has reflected on Bergson's famous division between space and duration, "Take a lump of sugar: it has a spatial configuration. But if we approach it from that angle, all we will ever grasp are differences in degree between that sugar and any another thing."[21] Via the spatial differences of degree (via gradients), an RNN, for example, generates the "attend" and "tell" outputs, deciding that "this is a woman on a sidewalk" because of its spatial distance from any other thing. Returning to Deleuze's account of Bergson's sugar cube: "But, it also has a duration, a rhythm of duration, a way of being in time that is at least partially revealed in the process of its dissolving, and that shows how this sugar differs in kind not only from other things, but first and foremost from itself. This alteration, which is one with the essence or substance of a thing, is what we grasp when we consider it in terms of Duration."[22]

It is possible to reimagine the attributes of something—"the essence or substance of a thing"—that the machine learning algorithm derives entirely from spatial clusters as durational qualities that differ in kind from themselves and others. The algorithm's generation of attributes carries only the lightness of likelihoods, of differences of degree. A cloud ethics freights the machine learning algorithm with the full durational weight—the waiting, the hesitation, the doubtfulness—of the essence or substance of a thing. In this way, the weight could be made to count as a meaningful moment of ethicopolitical decision. And so, when an algorithm for immigration decisions is said to weight the data inputs from past visa refusals, for example, the durational weight would surface a different kind of attribute, something unattributable. All the past moments of waiting in line (the wait in the weight), every past gathering of an incomplete visa application, all the persistent indeterminacies that differ in kind from themselves—these are the materials of a cloud ethics.

In revisiting the weights of algorithms, I am proposing that there are moments of undecidability within all technologies that enter the world as algorithmic decision systems. With the experience of undecidability comes also a mood of doubtful hesitation at the opening of the aperture.[23] Such hesitation is crucial to the forging of new ethicopolitical arrangements, because it expresses the weight of undecidability that is never eradicated entirely by the algorithm's output. I cannot know in advance the full consequences of what will be decided, but nonetheless a decision will be made. It seems that contemporary algorithmic decisions are precisely seeking to lighten the ethicopolitical load of undecidability. They must "feel the weight" of their decisions, which is a matter of not only weighting alternatives in a calculation but also weighing what is risked of ourselves and our relations to others in that calculus.[24]

With cloud ethics, what is condensed of the future by algorithms becomes something materially heavier and more burdensome, a kind of residue in the present. All algorithmic decisions contain within them the residue, or the sediment, of past political weightings. As we saw with the racialized policing and security algorithms with which I opened the introduction to this book, the condensed future ("he will pose a risk to the crowd") contains within it the residue of all the violence of past colonial histories, migrations, journeys, and border crossings, a fulsome sediment of all the actions and transactions of past movements in the name of justice. And in this residue one can trace the fork in the road that was not taken, the "trace of the rejected alternative" that continues to lodge itself within the cracks of the machine's capacity to learn.[25]

Opacity

Amid the widespread moral panic surrounding the black box of the algorithm and the political demands for transparency, accountability, and explainability, in this book I make a counter case for opacity and the giving of partial accounts. I have argued that responsibility for algorithmic decisions can never take the form of a clear-sighted account of the action. "Responsibility," as Thomas Keenan writes, "is not a moment of security or cognitive certainty" in which one makes a "choice between yes and no, this or that."[26] To establish a moment of cognitive certainty regarding the algorithm—to demand its transparency and accountability—is not at all the politics of a responsibility worthy of the name. Indeed, with its new forms of cognitive computing and cloud reasoning, as I discuss in chapter 1, the contemporary algorithm is well attuned to generating clear-sighted precise outputs from otherwise occluded situations. Contra the demand for transparency, then, the opacity of the scene, and the undecidability of its salient features and attributes, is the precondition for politics and ethics. "What could responsibility mean," asks Keenan, "without the risk of exposure to chance, without vulnerability," without the undecidability of what is taking place in a scene?[27] Running against the grain of algorithmic logics that show, attend, and tell the meaning of a situation, responsibility means that one cannot see a clear path ahead because the meaning of the situation is undecidable. The multiple branching points of the decision tree or random forest algorithm do in fact carry the risk of exposure to chance and vulnerability. A different kind of responsibility dwells at these forks in the path, bearing the opacity of future consequences of the weighting of one route over another.

In this book I differentiate the search for an encoded ethical framework for algorithms from an ethicopolitics that is already present within the algorithm's arrangements. In so doing, I draw a distinction between a *moral code* of prohibition and permission (what algorithms may or may not do) and *an ethics of the orientation one has to oneself and to others* (how algorithms are generating relations of self to self, and self to others).[28] This distinction follows Spinoza's broader philosophical concern that "ethics has nothing to do with a morality."[29] Our ethicopolitical relations do not follow strictly from some already formulated moral code, for we never have the security of these grounds from which to act. As Deleuze has reflected on Spinoza's ethics, "You do not know beforehand what good or bad you are capable of; you do not know beforehand what a body or a mind can do, in a given encounter, a given arrangement, a given combination."[30] Similarly, for Isabelle Stengers, "every time we use the term 'ethics' we must obviously distinguish it from the term 'morality,'" so that her speculative ambition is ethical in the sense of ethos or habit.[31]

Just as Spinoza's ethics are ever in formation, forged through ethos, encounters, arrangements, and combinations, so a cloud ethics is located in the encounters, arrangements, and combinations through which the algorithm is generated. Some of these encounters and arrangements involve human beings; others are between unsupervised algorithms and a corpus of data; and more still involve algorithms interacting with other algorithms. This is an ethics, then, that puts into question the authority of the knowing subject and opens onto the plural and distributed forms of the writing of algorithms. Understood in this way, "ethics is able to exceed its traditional determination" and be "displaced" and considered "anew" in the relations forged with other beings.[32] The possibility of what Derrida calls "the opening of another ethics" is a different kind of opening to the imagination of the prizing open of a black box or the illumination of a dark calculus with moral oversight.[33] It is an opening onto the edges and limits of an iterative writing that is profoundly political, each one of us drawn into the proximities of the output. Algorithms engage in writing in and through their dense relations with people and things in a data scene, their authorship uncertain and their emergence opaque. Here, in the opacity of not knowing in advance what the relations could be, we have found an opening for taking responsibility.

If an ethical response to algorithmic decisions is to begin from opacity and not from the demand for transparency, then what does this mean for the strategies of a cloud ethics? Where the political demands for transparency reassert the objective vision of a disembodied observer, to stay with opacity is to insist on the situated, embodied, and partial perspective of all forms of scientific knowledge.[34] Understood in Donna Haraway's terms, to confront opacity would not be a matter of failing to open the black box or to see something, but rather it would acknowledge that "the knowing self is partial in all its guises," that practices of visuality necessarily enroll the opacity of ourselves and others.[35] Far from a lack of transparency limiting accountability, the condition of opacity "allows us to become answerable for what we learn how to see."[36] This "we" who learns, as I discuss in the machine learning practices of chapter 2, is a composite figure, "stitched together imperfectly" from the iterative learning of humans and machines.[37]

In the chapters of this book, I have been concerned to show how machine learning algorithms generate their outputs through the opaque relations of selves to selves, and selves to others, and how these outputs then annul and disavow the opacity of their underlying relations in the name of a crystalline clarity of the output. One mode of resistance to the clarity of the algorithmic decision resides within the opacity of the subjects and what cannot be known of their relations. Let us consider a practical example of such a strategy. In the experimental development of the neural networks for show, attend, and tell, the algorithm designers explain that when their model makes a mistake, they are able to "exploit" what it manifests to gain "intuition into what the model saw."[38] This kind of seeing on behalf of the algorithm sustains the god's-eye vision of a computer scientist who may redistribute their way of seeing at will. A cloud ethics can stay with the difficulty of intuiting what the model saw as a form of partial, situated, and locally specific account. Among the examples of the mistakes used to productively generate new learning is "a woman holding a clock in her hand" and "a man wearing a hat and a hat on a skateboard." The algorithms have learned to recognize a clock and a hat, even when these are absent in the scene, via their exposure to a corpus of data that has supplied the feature vectors for these objects. The algorithm's relations with a contingent and opaque set of multiple other gradients have generated the mistaken presence of a clock and a hat (figures 6.4 and 6.5).

The CNNs have generated an output of attentiveness to the focus of the scene (shown in the images as an area of white light)—this output itself contingent on millions of past parameters—which enters the RNN as a data input. The RNN then generates the "tell," or the natural language description of the scene, assigning a probability for each word in the descriptor. In the (mis)recognition of "A man is talking on his cell phone while another man watches," the algorithm has assigned a probability of 1.00 (or a certainty) to the indefinite article A, for example, but probabilities of 0.46, 0.24, and 0.17 to the nouns man, cell, and phone. These likelihood scores—low and yet apparently the most probable—are entirely contingent on multiple algorithms' past relations with unknown other scenes. As I explain in chapter 4, the mistaken recognition of the algorithm is not merely error or errancy, for the output is entirely reasonable given the algorithm's relations with a data world. In terms of resistant practices, a cloud ethics is concerned not with fixing the algorithm or eradicating its errancy, but rather with attending to how these moments give accounts of algorithmic reason. What will take place when the recognition of a clock that is not present becomes the recognition of a weapon? What will be authorized by the most likely cell phone that is an absent presence in the scene?

To be clear, a cloud ethics will point to the moments when partial and situated algorithmic reason is sheltered as "mistake" or "error" and will reopen what is incomputable in the making of a likelihood score. Such a strategy is not the same as making a calculation transparent or accountable because it recognizes that the 0.14 assigned to child or 0.46 and 0.37 to border and fence are contingent on a vast multiplicity of necessarily opaque relations between other people, algorithms, things, data, gradients, features, and images. When the algorithm "tells" us that this is what matters in a scene, one could never have a transparent account of how the story has been made tellable. Instead, there are only the intransigent political relations that persist in the scene, the moments of nonclosure that reopen the problem and the politics.

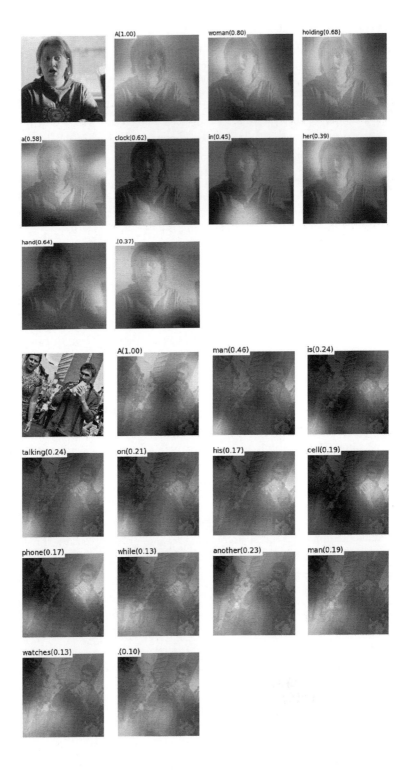

Contemporary machine learning algorithms operate with a specific concept of what it means to be together or to have things in common with others. In the ontology of algorithms, the mechanism that connects people, entities, and events is the attribute. The sphere of society and social relations thus exists for algorithms only as a series of groups, or *clusters*, defined by the spatial proximities and distances of their attributes. To govern a population through its attributes involves a rather different imagination of the relationship between the universal (e.g., shared values or moral principles, ideas of economy, common laws) and the particular (that which is particular to the subject as individual) than that which characterizes accounts of neoliberalism or biopolitics. The machine learning algorithm iteratively moves back and forth between the ground truth attributes of a known population (known in terms of the attribution of qualities) and the unknown feature vector that has not yet been encountered. In the space between a computer science universal of attributed feature space and the particularity of the new entity, the algorithms learn how to recognize and describe what is happening in the world. Thus, for example, the arrested person in the custody suite is rendered knowable by algorithms that calibrate their feature vector against the risk-scored attributes of another population—these encounters unrelated except via this new relation of attribution and adjudication.[39] Likewise, when a group of teenagers try to board a bus to attend a protest against the killing of a young African American man in their city, their political rights to gather with unknown others are annulled by algorithms that render them knowable via the clustering of their shared risk attributes.[40] What kind of relation to oneself and to others is manifest in this curious relation of known attributes to unknown others? How can the algorithm's dissolution of other forms of being together be resisted? How might a cloud ethics find political spaces for gatherings and collectivities that are precisely unattributable?

In the opening lines of this chapter, I cite Derrida's remarks on the ethicopolitical stakes involved in resisting apartheid. "This is a moment," he reflects, "of strategies, of rhetorics, of ethics, and of politics," a moment when it

Figures 6.4 and 6.5 Neural image caption generation with visual attention. The two examples display inaccurate outputs, with each word in the descriptor assigned a likelihood score. Xu et al., "Show, Attend and Tell," 14, 16.

is imperative to find the nonclosures at the limits of the border.[41] It seems that the struggles of our contemporary moment—resisting the rise of far-right nationalist groups, racist doctrines of anti-immigration, the everyday violences of sexism, racism, and bigotry—confront also the profound difficulties of finding an opening at the limit of the frame. Indeed, many of our most important and difficult historical political struggles—antiapartheid movements, the civil rights movement, campaigns for LGBTQ+ rights—would arguably have been impeded by state access to algorithms that could learn to attribute future threats. In short, machine learning algorithms are actively making it more difficult for new ethicopolitical claims—those claims not already registered as claimable—to be made in the world. In this context, a cloud ethics must be able to locate ways of being together that resist the algorithmic forces of attribution.

In an attributive society, what is perhaps most imperiled and at risk, and yet also most fugitive and precious to ethicopolitical life, is that which is *unattributable*. The unattributable—or a potentiality that cannot be attributed to a unitary subject—matters to ethics and politics because it does not locate its grounds in a clear *I* who makes a claim, but rather remains open to other claims not yet made. When machine learning algorithms segment a social scene, generating clusters of propensities, for example, to vote for a particular political party, or to respond to targeted social media, or to pose a high risk of recidivism, everything must be attributed, even the outliers understood as distant gradients from the curve of normality. The unattributable remains within the scene, however, exceeding the algorithm's capacity to show and tell, and opening onto a different kind of community and a different mode of being together. The attributive function that is the animating device of machine learning is, in fact, a more slippery and open device than its role in computer science might suggest. A cloud ethics will call this to mind each time it engages with the algorithm's showing and telling: the attribute is a cluster of features that remain persistently unattributable to any subject as such. There is a gap, a breach, an opening between the output of the algorithm and the body of the subject in whom it becomes actualized, at the border, in the police custody suite, at the employment center, on the city street.

The resistant orientation to attribution I have in mind does not stand outside or in opposition to the attributive power of the algorithm. In common with all other forms of resistance, this is an orientation that shares the same horizon as the mode of power. Indeed, I have explained in this book the extent to which human subjects dwell within the algorithm, as it comes into being also through relations with subjects and objects. What is necessary is not a resistant overturning of the logics of algorithm but a fuller and more amplified attunement to the ethicopolitics that are always already present in algorithmic reason. Thus, I envisage a cloud ethics that takes seriously the governmental form of the attribute and dwells with it. Yet, a cloud ethics must also refuse the attachment of attributes to entities in favor of the immanence of all forms of writing and attribution. As Deleuze has captured the attribute, following a Spinozan ethics, it expresses "an essence that is an unlimited, infinite quality" so that a "single substance" has an "infinity of attributes."[42] The unlimited and infinite quality of the attribute offers a resistant breach in the algorithm's society of attributes, for the algorithm's attribute must always be a finite and fixed essence to be classifiable and recognizable. A cloud ethics recognizes, in contrast, the unlimited and infinite emergence of attributes that will defy the boundaries of clusters and feature vectors. Put simply, the attribute is unattributable to those on whom its effects are exerted.

There is a political refusal involved in claiming the unattributable. It is a refusal to be governed by the attribute, an affirmation of "how not to be governed *like that*," as Foucault has described it, a certain "will not to be governed" as a cluster of attributes.[43] My ethicopolitical relations to myself and to others, then, cannot be limited to recognizable attributes but must begin from a recognition of singularity. As Simon Critchley writes, "Ethics is first and foremost a respect for the concrete particularity of the other person in his or her singularity."[44] This means that the other person cannot simply be an "example" of some set of finite qualities derived from the gradients of norm and anomaly of others. "Ethics begins as a relation with a singular, other person," continues Critchley, this singular other person who then "calls me into question" so that "politics begins with ethics."[45] As the algorithm presents us with an attributive other who is never singular or particular, it is crucial that we refuse the paradigm of attribution and amplify the unattributable. I cannot know the essence of the other as a cluster of attributes, and when the algorithm presents me with the likelihood of their attribution, I must begin again from their singularity.

The question of an ethics of the unattributable is not only a matter of thought or of the philosophy of algorithms in society. It must also be a matter of the practice of critique and the situated struggle for alternative routes that are foreclosed in the calculation of an output. There have been many occasions when I have asked the same question of border and policing authorities, newly equipped with their predictive recognition algorithms: "How will you know the characteristics of this unknown person who appears?" Always, the response has been some variant on the same notion: "We do not need to know them in particular; we have seen their attributes before." The unattributable is thus at the very heart of the struggles over the place of algorithms in our societies. A cloud ethics must pursue a conception of being together as a society of the unattributable, where what we share in common is precisely our singularities, and where our encounters with others can never be limited to a finite collection of attributes.[46] All the many moments when algorithms gather to cluster our attributes, to assemble against us, when they seek to silence the claim before it could even be made—these are the moments that must remain open to what cannot be attributed. Ethicopolitical life is about irresolvable struggles, intransigence, duress, and opacity, and it must continue to be so if a future possibility for politics is not to be eclipsed by the output signals of algorithms.

NOTES

Introduction

1 Observations of tech start-ups' pitches for government procurement of algorithmic systems, London, March 2016. An earlier iteration of the algorithms is described in a 2012 video, "Monitoring Protests from Unstructured Text."

2 Vice News journalist Jason Leopold made a Freedom of Information Act request seeking all records pertaining to Freddie Gray and the subsequent protests and riots in Baltimore, Maryland. For the full disclosed documents, see "DHS FOIA Documents Baltimore Protests Freddie Gray."

3 For the text of the Geofeedia case study, see ACLU, "Baltimore County Police Department and Geofeedia Partner."

4 Computer scientists give accounts of how they use samples of training data to train their model to recognize and extract text when it appears in an image. See, for example, Misra, Swain, and Mantri, "Text Extraction and Recognition," 13–19.

5 Browne, *Dark Matters*, 26.

6 Laughland, "Baltimore Unrest."

7 Though many specific algorithms become secret and proprietary when bought and licensed by corporations and governments, the upstream science of their genesis is much more public, incremental, and debated. Methodologically, a degree of speculation is necessarily involved in identifying which broad family of algorithms is being used in a particular application. Among the many classic computer science papers on contemporary machine learning algorithms I have used to infer the operations of "secret" or "proprietary" algorithms are LeCun, Bengio, and Hinton, "Deep Learning," 436–44; Hinton and Salakhutdinov, "Reducing the Dimensionality of Data," 504–7; Krizhevsky, Sustskever, and Hinton, "ImageNet Classification"; Simonyan and Zisserman, "Very Deep Convolutional Networks"; and Hinton, "Deep Neural Networks for Acoustic Modeling," 82–97.

8 "Amid the uncertainties, potentialities, and dangers created by the regime of computation," writes N. Katherine Hayles, "simulations—computational and narrative—can serve as potent resources with which to explore and understand the entanglement of language with code . . . and subjectivity with computation." Hayles, *My Mother Was a Computer*, 242.

9 Clustering techniques in machine learning are experimental and exploratory in the sense that they discover previously unknown structure in a large volume of data. The algorithms iteratively build a model by grouping the data points that are most similar in their attributes. The model is then used to predict likely behavior in the future, such as the propensity for a group to "churn," or to switch commercial providers of, say, energy, credit, insurance, or telecommunications. The aim in clustering, as Ethem Alpaydin describes, "is to find clusters of input" data and to define each group as "similar in their attributes." Alpaydin, *Machine*

Learning, 112. A set of algorithms designed for customer segmentation models are also widely used for security and policing. For example, computer scientists have trained machine learning algorithms on Twitter data to determine the attributes of support for the Islamic State, training their classifier to "predict future support or opposition of ISIS [Islamic State of Iraq and Syria] with 87% accuracy." Emerging Technology from the arXiv, "Twitter Data Mining"; Caruso, "Can a Social Media Algorithm Predict a Terror Attack?"

10 Crawford, "Artificial Intelligence's White Guy Problem." See also John Markoff's interview with Crawford on the ethical conduct of the designers of machine learning algorithms; Markoff, "Artificial Intelligence Is Far from Matching Humans." See also Metcalf, Keller, and boyd's report on the ethical design of big data analysis, "Perspectives on Big Data."

11 O'Neil, *Weapons of Math Destruction*.

12 Pasquale, *Black Box Society*.

13 In Michel Foucault's 1981 series of lectures at the Catholic University of Louvain, *Mal faire, dire vrai [Wrong-Doing, Truth-Telling]*, he explains that his interest lies not in establishing the basis or grounds of truth but instead in "the forms of the enterprise of truth-telling" of particular scientific or technical claims. He proposes a counterpoint to positivist claims in what he calls "astonishment before the proliferation of truth-telling, and the dispersal of regimes of veridiction." This proliferation of truth telling by algorithms, and the regimes of veridiction that make it possible, is of interest to me in this book. Foucault, *Wrong-Doing, Truth-Telling*, 20–21.

14 Cheney-Lippold, *We Are Data*, 9.

15 Connolly, "Beyond Good and Evil," 366.

16 Michel Foucault, in addressing his genealogy of ethics, depicts a "great change" from Greek society to Christian society dwelling "not in the code but in the ethics," which is "their relations to themselves and to others." Foucault, "On the Genealogy of Ethics," 253–80. A related distinction between ethics as code and ethics as practice can be found in the work of Gilles Deleuze, for whom the term *morality* denotes a set of rules or codes, whereas *ethics* is immanent to a "mode of existence," or a form of life. Deleuze, *Negotiations*, 100; and Deleuze, "On the Difference between the *Ethics* and a Morality," 17–28.

17 The "cloud layer," as Bratton describes it, forms "the second from the bottom of the Stack," so that "cloud platforms may structure Cloud polities." Bratton, *The Stack*, 9, 369. See also Hu, *Prehistory of the Cloud*.

18 Peters, *Marvelous Clouds*, 2. Media theorist Jussi Parikka similarly directs attention to the material and geophysical making of media; Parikka, *Geology of Media*.

19 Butler, *Giving an Account of Oneself*, 12.

20 Butler, *Giving an Account of Oneself*, 17.

21 Amaro, "Race, Surveillance."

22 Mathematician A. A. Markov defined the ideal algorithm as one that offered "precision," "definiteness," and the "conclusiveness" of a desired result. Markov, *Theory of Algorithms*, 1.

23 Wittgenstein, *On Certainty*, §653–57.

24 Wittgenstein, *Philosophical Investigations*, §136. I am indebted to Jeremy Schmidt for conversations on Wittgenstein and mathematics as praxis.

25 Wittgenstein, *On Certainty*, §655.

26 Wittgenstein, *On Certainty*, §651.

27 As Karen Barad has argued compellingly, "It is not merely the case that human concepts are embodied in apparatuses, but rather that apparatuses are discursive practices . . . through which 'objects' and 'subjects' are produced." In Barad's reading, scientific apparatuses are agents that enact boundaries, becoming the condition of possibility of claims to cause and effect. Barad, *Meeting the Universe Halfway*, 148, 175.

28 Foucault, "Ethics of the Concern for Self," 299.

29 Seaver, "Algorithms as Culture," 2.

30 Among the many depictions of the algorithm as a sequence of steps, for Tarleton Gillespie, an algorithm is "a recipe composed in programmable steps." Gillespie "Algorithm," 19. Pedro Domingos defines an algorithm as "a sequence of instructions." Domingos, *Master Algorithm*, 1. David Berlinski identifies "an algorithm [as] a finite procedure." Berlinski, *Advent of the Algorithm*, xviii.

31 DeLanda, *Intensive Science and Virtual Philosophy*.

32 David Berry has distinguished between different algorithmic forms to show their different implications for ordering and acting on society. Berry, *Philosophy of Software*.

33 In Noel Sharkey's assessment of the different juridical options for the control of autonomous weapons systems, he writes that "we must urgently begin to develop a principle for the human control of weapons that is founded on an understanding of the process of human reasoning." For Sharkey, the "reframing" of "autonomy in terms of human control" has the effect of making it "clear who is in control, where and when." Sharkey "Staying in the Loop," 23–38.

34 DeLanda, *Philosophy and Simulation*, 44–45.

35 For Totaro and Ninno the recursive logic defines "the logic of algorithms" and is the "specific form in which the concept of function occurs in everyday life." Totaro and Ninno, "Concept of Algorithm," 32. See also Beer, "Power through the Algorithm?," 985–1002.

36 The current state-of-the-art design of neural networks for voice and image recognition, for example, is experimenting with teaching algorithms how to teach themselves which hidden layers could be bypassed without significantly altering the output. LeCun, Bengio, and Hinton, "Deep Learning."

37 So-called risk assessment engines powered by machine learning algorithms are being used by judges to predict the likely probability of a convicted person reoffending if they return to the community. Research has shown that the algorithms are overestimating the risk of future crimes by black defendants—and therefore proposing prison as the optimal judgment—even where race is not an input datum in the algorithm. See Angwin et al., "Machine Bias"; and Liptak, "Sent to Prison." The United Kingdom's Durham Police Constabulary deploys the Harm

Assessment Risk Tool (HART) to assess the likelihood of an individual reoffending if they are granted bail. See Bridge and Swerling, "Bail or Jail?"

38 Mackenzie, *Machine Learners*, 27.

39 Parisi, *Contagious Architecture*, 2.

40 Winslett, "Rakesh Agrawal Speaks Out."

41 Amoore, *Politics of Possibility*, 50–52, 139–41.

42 Turing, "Systems of Logic Based on Ordinals," 161–228. In letters sent to Max Newman in 1940, Turing is "not sure whether my use of the word 'intuition' is right or whether your 'inspiration' would be better" and considers that Newman is "using 'inspiration' to cover what I call 'ingenuity.'" Turing, "Letters on Logic to Max Newman," 212.

43 Diamond, *Wittgenstein's Lectures*, 150, 251.

44 Devlin, *Math Gene*, 110.

45 Stengers, *Thinking with Whitehead*, 413.

46 Intelligence and Security Committee of Parliament, *Privacy and Security*, 8.

47 Schneier, "NSA Robots."

48 O. Halpern, *Beautiful Data*, 21.

49 Crary, *Suspensions of Perception*, 2.

50 Kittler, *Discourse Networks 1800/1900*; Kittler, *Literature, Media, Information Systems*.

51 "The whole difficulty of the problem," writes Bergson, "comes from the fact that we imagine perception to be a kind of photographic view of things, taken from a fixed point by that special apparatus which is called an organ of perception." Bergson, *Matter and Memory*, 31.

52 Bergson, *Creative Mind*, 120.

53 Bergson, *Matter and Memory*, 34.

54 Interview with designer of algorithms for the extraction and analysis of anomalous objects from video data, northwest England, June 2014.

55 As Lisa Parks and Caren Kaplan write, "Drones are not idle machines hovering above; they are loaded with certain assumptions and ideologies . . . they are also ideas, designs, visions, plans, and strategies that affect civilians on the ground." The assumptions and weights of deep neural network algorithms form a central element of just such a loading of ideas and design. Parks and Kaplan, *Life in the Age of Drone Warfare*, 9. For a compelling account of the embodied and durational experience of drone technologies, see also Wilcox, "Embodying Algorithmic War," 11–28.

56 Hansen, *New Philosophy for New Media*, 104.

57 Deleuze, *Bergsonism*, 32.

58 Eubanks, *Automating Inequality*.

59 *Nature* Editors, "More Accountability for Big Data Algorithms."

60 Brandom, "New System Can Measure the Hidden Bias."

61 Foucault, "What Is an Author?," 207.

62 Keenan, *Fables of Responsibility*, 3.

63 Keenan, *Fables of Responsibility*, 51.

64 In N. Katherine Hayles's study of the forms of nonconscious cognition distrib-

uted across human and technical systems, she argues that "nonconscious cognitions in biological organisms and technical systems share certain structural and functional similarities, specifically in building up layers of interactions." Hayles, *Unthought*, 13.

65 Haraway, *Staying with the Trouble*, 3–4.

66 Butler, *Giving an Account of Oneself*, 18, 29.

67 Haraway, "Situated Knowledges," 583.

68 Hayles, *How We Became Posthuman*, 287.

69 "If we understand machine learning as a data practice that reconfigures local centers of power and knowledge," writes Adrian Mackenzie, "then differences associated with machine learners in the production of knowledge should be a focus of attention." Mackenzie, *Machine Learners*, 10.

1. The Cloud Chambers

1 C. T. R. Wilson, "On the Cloud Method," 194.

2 C. T. R. Wilson, "On the Cloud Method," 195.

3 C. T. R. Wilson, "On the Cloud Method," 196.

4 Galison, *Image and Logic*, 140.

5 C. T. R. Wilson, "On the Cloud Method", 199.

6 Dodge, "Seeing Inside the Cloud"; see also Dodge and Kitchin, *Mapping Cyberspace*.

7 Konkel, "The CIA's Deal with Amazon," 2.

8 The data of the seventeen ICITE agencies include information from sensors, satellites, UAV images, open-source social media, internet image files, text, video, and voiceover internet files. For further discussion of how these different forms of structured and unstructured data are ingested and analyzed, see Amoore and Piotukh, "Life beyond Big Data."

9 Konkel, "The CIA's Deal with Amazon," 2.

10 Central Intelligence Agency, "CIA Creates a Cloud."

11 Alpers, "The Studio, the Laboratory," 415.

12 Hayes, "Cloud Computing," 9.

13 boyd and Crawford, "Critical Questions for Big Data," 662–79; Mayer-Schönberger and Cukier, *Big Data*; and Kitchin, *Data Revolution*.

14 McCarthy, "Architects of the Information Society," 2.

15 Human geographer Sam Kinsley has urged greater attention to the materiality of the virtual and, specifically, to the details of where and how virtual geographies become actualized. Kinsley, "Matter of Virtual Geographies," 364–84.

16 Bratton, *The Stack*, 29.

17 O'Brien, *Definitive Guide to the Data Lake*.

18 Jaeger, "Where Is the Cloud?," 4.

19 Though Sun Microsystems brought to market the first mobile and modular data centers inside shipping containers, Google later patented designs for floating data centers of shipping container servers stacked within cargo ships. Appealing to the capacity of the floating data center to respond to emergencies such as the Fu-

kushima earthquake, and to store and analyze data in international waters, the Google cloud enters the geopolitics of logistics and exceptional spaces.

20 European Commission, "What Does the Commission Mean?"

21 Intelligence and Security Committee of Parliament, "Transcript of Evidence," with my additions from video of proceedings.

22 Crawford and Schultz, "Big Data and Data Protection," 99, 105.

23 Crawford and Schultz, "Big Data and Data Protection," 110.

24 Paglen, "Trevor Paglen Offers a Glimpse."

25 Paglen, *Invisible*.

26 Galison, *Image and Logic*, 72.

27 Paglen, "Trevor Paglen Offers a Glimpse," 4; my emphasis.

28 Bridle, *New Dark Age*, 8.

29 Crary, 24/7; Rose, "Rethinking the Geographies"; and Mattern, *Code and Clay, Data and Dirt*.

30 The design of algorithms for a national border-control system was outsourced through a layered series of small contractors and start-ups. The major software company that held the government contract had no working understanding of how the algorithms were interacting across databases. In this case, a "place to look" was a nondescript rented office in North London, hosting a small group working on the design, validation, and testing of a series of interacting algorithms. See Amoore, "Data Derivatives," 24–43.

31 In Luciana Parisi's description of how the spatial outline of an object is overflowed by digital computational models, she notes, "What is at stake with these generative algorithms is that the notion of discreteness has changed, and now includes a model of interactive agents that evolve in and through time." Understood in these terms, there could never be a definitive building or object in which to locate cloud computation, for the forms of these places would be (computationally) malleable and evolving. Parisi, *Contagious Architecture*, 45.

32 Graham, *Vertical*, 66.

33 Scorer and Wexler, *Cloud Studies in Colour*.

34 Haraway, "Situated Knowledges," 581.

35 Donna Haraway traces the etymology of "trouble" to its French roots of "to stir up," "to make cloudy," or "to disturb." Writing against the notion of trouble as something to be averted through security and safety, Haraway develops a form of "staying with the trouble" that learns how to "be truly present" in the "myriad unfinished configurations." Haraway, *Staying with the Trouble*, 1; and Haraway, "Situated Knowledges," 583. Understood in this way, to pursue a cloud ethics is also to disturb the sediment of algorithmic logics, to make cloudy the otherwise crystalline definiteness of the output.

36 Galison, *Image and Logic*, 97.

37 Galison, *Image and Logic*, 67.

38 John Durham Peters proposes that digital media extend historical distributed infrastructures of the environment as media, so that "media are perhaps more interesting when they reveal what defies materialization." Peters, *Marvelous Clouds*,

11. Similarly, for Derek McCormack, techniques of remote sensing are better understood as "sensing spectrality" rather than "a project of techno-scientific mastery." McCormack, "Remotely Sensing Affective Afterlives," 650.

39 Barad, *Meeting the Universe Halfway*, 176.

40 Erickson et al., *How Reason Almost Lost Its Mind*, 9, 29.

41 Bergson, *Matter and Memory*, 31.

42 Digital Reasoning, *Synthesys Mission Analytics*, 4. See also "Machine Learning for Cybersecurity."

43 Leopold, "CIA Embraces Cloudera Data Hub."

44 For Hayles, cognizers perform as "actors embedded in cognitive assemblages with moral and ethical implications." Hayles, *Unthought*, 31.

45 Leopold, "CIA Embraces Cloudera Data Hub," 2.

46 Galison, *Image and Logic*, 67. Office of the Director of National Intelligence, *Intelligence Community*.

47 C. Anderson, "The End of Theory."

48 J. G. Wilson, *Principles of Cloud Chamber Technique*, 3.

49 Gentner, Maier-Leibnitz, and Bothe, *Atlas of Typical Expansion Chamber Photographs*, 11.

50 Foucault, *Security, Territory, Population*, 6.

51 Barad, *Meeting the Universe Halfway*, 148.

52 Elden, "Secure the Volume," 35–51; Crampton, "Cartographic Calculations of Territory," 92–103; Weizman, "Politics of Verticality."

53 O'Brien, *Definitive Guide to the Data Lake*, 4.

54 The data lake architecture underpins the ability of commercial and government agencies to gather and analyze "vast amounts of social networking platforms emerging data" in a "single, centralized repository." O'Brien, *Definitive Guide to the Data Lake*, 3.

55 Parisi, *Contagious Architecture*, 2.

56 Josephson and Josephson, *Abductive Inference*, 12.

57 Konkel, "The CIA's Deal with Amazon."

58 Dunning and Friedman, *Practical Machine Learning*, 23.

59 Interview with designers of CNN algorithms for facial recognition, London, May 2017.

60 J. G. Wilson, *Principles of Cloud Chamber Technique*, 122.

61 Rochester and Wilson, *Cloud Chamber Photographs*, vii.

62 Galison, *Image and Logic*, 130.

63 In the past, the use of data-mining techniques for security and intelligence was limited, in part, by the lack of available training datasets. The storage of data in the cloud provides a readily available supply of training data (and a flexible computational capacity) through which humans and algorithms can advance their capacities to perceive new emergent events.

64 Interview with UK government procurement officers, London, October 2017.

65 Office of the Director of National Intelligence, *Intelligence Community*, ii.

66 Hayles, *How We Think*, 230.

67 Office of the Director of National Intelligence, "ICITE Fact Sheet."

68 "Web Intelligence"; my emphasis.

69 O. Halpern, *Beautiful Data*, 16, 40.

70 Derrida, *Archive Fever*, 17.

71 James Bridle's interactive installation is available at Hyper-Stacks, accessed June 2016, http://hyper-stacks.com.

72 Derrida, *Archive Fever*, 18.

73 Observations conducted in the Department of Physics, Durham University, August 2015.

74 Interviews and observations with the designers of credit card fraud detection algorithms, Frankfurt, April 2014.

75 Stengers, *Cosmopolitics I*, 22.

76 Bal, "Commitment to Look," 145–62; and W. J. T. Mitchell, "There Are No Visual Media," 257–66.

77 Halpern, *Beautiful Data*, 156.

78 Stengers, *Cosmopolitics I*, 31.

79 D. Mitchell, *Cloud Atlas*, 389.

2. The Learning Machines

1 Newman et al., "Can Automatic Calculating Machines Be Said to Think?," 10.

2 Three years earlier, Jefferson had delivered his famous lecture "The Mind of Mechanical Man" at the Royal College of Surgeons. In the lecture he had reflected that neuroscience was being "invaded by the physicists and mathematicians," so that "we are being pushed to accept the great likeness between the actions of electronic machines and those of the nervous system." Jefferson, "Mind of Mechanical Man," 1105–10. For Jefferson, distinct and important differences between human and machine persisted, not least that the human brain could sustain "vast cell losses without serious loss of memory," and that the machine "can answer only problems given to it, prearranged by its operator" (1109). During this postwar era of twinned cybernetics and Cold War rationalities, the neurosurgeon's world of the material structures of the brain ran against the grain of a belief that "machines might outdo humans in executing algorithms." Erickson et al., *How Reason Almost Lost Its Mind*, 9.

3 "Can Automatic Calculating Machines Be Said to Think?," 11; my emphasis.

4 Turing, "Systems of Logic Based on Ordinals," 192.

5 "Can Automatic Calculating Machines Be Said to Think?," 13.

6 John Cheney-Lippold's argument that "we are data" emphasizes how we humans are "temporary members of different emergent categories," these categories having been assigned meaning "without our direct participation." I am less persuaded that humans do not have direct participation in emergent forms of classification and categorization, and I wish to attend to the entangled learning that situates humans within algorithms and algorithms within humans. Cheney-Lippold, *We Are Data*, 5.

7 Anne Marie Mol's ethnographic study of one disease—atherosclerosis—suggests

how the multiplicity of the disease is rendered coherent, or made to cohere, through the plural practices of imaging, scientific discussion and disagreement, and everyday negotiations. Mol, *Body Multiple*. This making of a cohering whole from multiplicity also characterizes the actions of algorithms. Brian Rotman notes that digital data environments are generating "a network 'I'" who is "plural and distributed," "heterogeneous and multiple." Rotman, *Becoming Beside Ourselves*, 8.

8 A *feature space* is a mapping of input data conducted by processing subsets of data so that "after the entire input is processed, significant features correspond to denser regions in the feature space." These denser regions are "clusters." Comaniciu and Meer, "Mean Shift," 605.

9 Lin et al., "Automatic Detection," 802–10.

10 In machine learning and pattern recognition applications, a feature vector is a means of numerically representing an n-dimensional object in space so that it can be computed. Thus, for example, the image of a human face is rendered computable in a biometrics application via the extraction of feature vectors corresponding to pixelated data points. In the movement of the hands of surgeons and surgical robots, the feature vectors denote the position along a trajectory in three-dimensional space.

11 Lin et al., "Automatic Detection," 806.

12 Reiley, Plaku, and Hager, "Motion Generation of Robotic Surgical Tasks."

13 N. Katherine Hayles describes a "cognitive assemblage" of human and technical "cognizers" who are "flexibly attending to new situations," incorporating modes of knowledge, and "evolving through experience to create new responses." Hayles, *Unthought*, 32. Adrian Mackenzie proposes that the "various subject positions" of machine learning are "neither unified nor fixed" but malleable and transformative. Mackenzie, *Machine Learners*, 207.

14 Van den Berg et al., "Superhuman Performance of Surgical Tasks."

15 Robotic surgery actively *premediates* the actualized surgery in the sense intended by Richard Grusin in his work on premediation. The robot's virtual architecture allows the surgeon to explore the trajectories of the surgery in advance of the actual surgery, anticipating the optimal trajectories of movement, experimenting, and preempting potential problems. Grusin, *Premediation*.

16 Prentice, *Bodies in Formation*, 227.

17 Prentice, *Bodies in Formation*, 227–29.

18 Prentice, *Bodies in Formation*, 231.

19 Popescu and Mastorakis, "Simulation of da Vinci Surgical Robot," 141.

20 Interview with obstetric surgical team, Newcastle-upon-Tyne, April 2016.

21 Barad, *Meeting the Universe Halfway*, 208.

22 Jochum and Goldberg, "Cultivating the Uncanny."

23 Following Henri Bergson, the cut and the multiplicity are actually two distinct forms of multiplicity. As Gilles Deleuze writes, "Remember that Bergson opposed two types of multiplicity—actual multiplicities that are numerical and discontinuous and virtual multiplicities that are continuous and qualitative." Deleuze,

Bergsonism, 79. For the singular cut to take place, the virtual multiplicities of past encounters between people and algorithms must be rendered as actual divisible numerical multiplicities, or feature spaces.

24 In a computer science paper on machine learning for robot surgery, the authors present their cognate findings in military robots, landmine detection, and robot inspection of nuclear contamination. The development of machine learning algorithms takes place indifferent to the environment as such because it understands the environment explicitly in terms of feature space and not place. Popescu and Mastorakis, "Simulation of da Vinci Surgical Robot," 137.

25 Hayles, *How We Became Posthuman*, 4.

26 Hayles, *How We Became Posthuman*, 287.

27 As Kant writes, "The *I think* must *be able* to accompany all my representations," so that "the thought that these representations given in intuition all together belong *to me* means, accordingly, the same as that I unite them in a self-consciousness." The reasoning subject is the anchor that secures all representations, all scientific and political claims, so that one conceives of oneself as a unity of thought and action. Kant, *Critique of Pure Reason*, 247.

28 In 2016, Intuitive Surgical reserved $100 million to settle an undisclosed number of claims for damages. The company's insurer, Illinois Union, sued Intuitive Surgical for its alleged failure to disclose the extent of medical injuries associated with its robot systems. Mehotra, "Maker of Surgical Robot."

29 As Noel Sharkey has proposed in his accounts of the human control of robot weapons, "There will be a human in the loop for lethality decisions." For Sharkey, the human in the loop is understood to be the locus of "deliberative human control," where space and time in the kill chain is afforded for "deliberate reasoning." Sharkey, "Staying in the Loop," 25, 35. See also Strawser, *Killing by Remote Control*.

30 Lucy Suchman and Jutta Weber detail the limits of the notion of ethics in relation to autonomous weapons systems. Reflecting on the iterative relationship between military robots, their machine learning algorithms, and the humans conducting postprocessing, they argue that "robot behaviour shifts to another level: it is no longer totally pre-programmed but, instead, more flexible. The behaviour is mediated by random effects of the system's architecture or learning algorithms, which can result in emergent effects which are then exploited postprocessing." Suchman and Weber, "Human-Machine Autonomies," 88.

31 Haraway offers a "method of tracing" or "following a thread in the dark" that stands in contrast to the illuminations sought by calls for transparency or an opening of closed systems. A Haraway-inspired tracing does not seek out the concealed origins of science, but rather the tracing makes material the threads that have been necessary to science's story of origins and objectivity. To follow a thread or a trace is thus not comparable to an opening of scientific black boxes, for it does not seek a moment of revelation and exposure. She describes a "touching silliness about technofixes (or techno-apocalypses)" and counters this with a determination to "follow the threads where they lead" in "real and particular places and times." Haraway, *Staying with the Trouble*, 3.

32 Interviews and observations with algorithm designers for a national border and immigration control system, July 2016.

33 I am grateful to Phil Garnett for pointing out the specificity of the notion of "good enough" in the practice of machine learning and generously discussing the potential consequences.

34 Dunning and Friedman, *Practical Machine Learning*, 14.

35 Dunning and Friedman, *Practical Machine Learning*, 16; my emphasis.

36 There is extremely important work being done to address how racism, sexism, and economic inequalities are becoming inscribed into the actions of algorithms. See Noble, *Algorithms of Oppression*; and Eubanks, *Automating Inequality*.

37 Ramon Amaro's work has focused on the capacities of the algorithm to recognize blackness and the new regimes of racialized recognition that accompany algorithmic techniques. Amaro, "Pre-emptive Citizenship."

38 Bucher, *If... Then*, 4.

39 Domingos, *Master Algorithm*, 189.

40 Quinlan, "Induction of Decision Trees," 82.

41 Quinlan, "Induction of Decision Trees," 83.

42 Quinlan, "Induction of Decision Trees," 104.

43 Though it may appear inconsequential that an algorithm may be less able to identify shape or outline than it is able to detect texture or color, in fact it is becoming increasingly important to our contemporary regime of recognition. For example, neural networks used in the assessment of medical images are more attuned to the texture or color of a human organ than they are to its shape or indeed to the contours of a lesion. Hendrycks et al., "Natural Adversarial Examples," 2.

44 LeCun, Bengio, and Hinton, "Deep Learning," 436.

45 Krizhevsky, Sutskever, and Hinton, "ImageNet Classification," 1097–1105.

46 Across a wide range of recognition algorithms, including tumor detection in medical imaging and facial recognition, the AlexNet algorithm is used as the pretrained model, with the domain-specific images (e.g., datasets of surgical photographs) added only to the training of the last hidden layer of the neural network.

47 Krizhevsky, Sutskever, and Hinton, "ImageNet Classification," 1097.

48 Amazon Mechanical Turk has become a vehicle for the human labeling of images that will be used to train a machine learning algorithm. Described by Amazon as "a marketplace for work that requires human intelligence," the casualization and precarity of this form of work means that a population of workers is engaged in the preparation of data for algorithmic systems that may at some future moment be used against them or others in their community.

49 Observations of the process of training algorithms for image recognition, Gothenburg, November 2016.

50 MacKay, *Information Theory*, 345.

51 Interviews and observations with designers of algorithms for airport border security, September 2016.

52 LeCun, Bengio, and Hinton, "Deep Learning," 436.

53 McGeoch, *Guide to Experimental Algorithmics*, 87.

54 Interview with test computer scientist working with IBM on facial recognition, London, June 2018.

55 All forms of recognition, as political claims, carry within them the possibility of misrecognition or being unrecognized. As Judith Butler has put the problem, "To ask for recognition, or to offer it, is precisely not to ask for recognition for what one already is. It is to solicit a becoming, to petition the future always in relation to the Other." Butler, *Precarious Life*, 44.

56 The Point Cloud Library is an open source repository of point cloud 2D and 3D images and object recognition algorithms. The Point Cloud Library offers techniques to "stitch 3D point clouds together" and to "segment relevant parts of a scene." As laser-scanning techniques generate datasets of 3D objects, from architects' scans of buildings to MRI scans of tumors, the 3D point clouds become training data for future neural network algorithms. See "About."

57 The algorithmic capacity to recognize 3D objects from multiple angles and viewpoints has also become the condition of possibility for facial recognition systems that can be deployed on the variegated and oblique angles of the CCTV camera surveillance architecture.

58 Mahler et al., "Dex-Net 1.0."

59 Mahler et al., "Dex-Net 1.0," 1.

60 Mahler et al., "Dex-Net 1.0," 2.

61 Interview with algorithm design team working on surgical robotics, detection of problem gambling, and biometrics recognition systems, London, September 2014.

62 Algorithm design team interview, London, September 2014.

63 Reflecting on his experiences of the events of May 1968, Foucault expresses how a set of political claims can be brought that have never previously been registered as claimable rights or encoded norms. Such claims that are "not part of the statutory domain" included "questions about women, about relations between the sexes, about medicine, about mental illness, about the environment, about minorities, about delinquency." In the bringing of these political questions, Foucault does not identify an appeal to political rights but rather a "liberation of the act of questioning" from existing political doctrines. One can imagine a bringing of questions that are not traditionally part of the statutory domain of politics as one element of a cloud ethics. Foucault, "Polemics, Politics, Problematizations," 115.

3. The Uncertain Author

1 Kirchner, "New York City Moves."

2 ProPublica's investigation obtained the risk scores assigned to seven thousand people arrested in Broward County, Florida, in 2013 and 2014 and analyzed the reoffending rate. Their study found that the algorithm used to predict the risk of reoffending was of not greater predictive value than a coin toss, but that the effects were acutely divided along racialized lines. "The formula was particularly

likely to falsely flag black defendants as future criminals, wrongly labeling them this way at almost twice the rate of white defendants," the ProPublica report concluded. Angwin et al., "Machine Bias."

3　The Council for Big Data, Ethics, and Society calls for a public discussion of how the ethical conduct of algorithm designers could be enshrined in a program of training and regulation. Metcalf, Keller, and boyd, "Perspectives on Big Data."

4　An influential editorial in *Nature* proposed that algorithmic bias could be scrutinized through the disclosure of the data sources used by the designers. The appeal to a unified body of data and a known designer similarly invokes the line of accountability to an authorial source. *Nature* Editors, "More Accountability."

5　For Foucault the author function does not consist merely of the attribution of a work to an individual, but rather "constructs a certain being of reason that we call 'author.'" The author function unifies the oeuvre, or the body of work, so that unresolved tensions or contradictions appear resolvable and "incompatible elements are at last tied together." Foucault, "What Is an Author?," 384–85.

6　Foucault, "What Is an Author?," 377.

7　Lev Manovich has depicted the distinction between narrative and database as one of competing forms, even as "enemies." Manovich, *Language of New Media*, 228. Contra Manovich, N. Katherine Hayles develops an understanding of the interdependent mediations of narrative and database, nonetheless proposing that "they remain different species, like bird and water buffalo." Hayles, *How We Think*, 178. With contemporary machine learning algorithms, however, the indeterminacy that Hayles suggests "databases find difficult to tolerate" moves closer to the "inexplicable" and "ineffable" she associates with the narrative form (178–79).

8　Kirby, *Quantum Anthropologies*, xi.

9　Reflecting on the Greco-Roman practice of writing the hupomnemata, or individual notebooks, Foucault describes writing that "resists scattering" by the "selecting of heterogeneous elements." The writing becomes a "matter of unifying heterogeneous fragments" into a body that is recognizable not as a "body of doctrine" but as a "certain way of manifesting oneself to oneself and to others." Foucault, "Self Writing," 212–13, 216.

10　The computational practice of natural language processing is historically intertwined with linguistic theory and, specifically, the debates on how language is learned by humans. Noam Chomsky's Platonic approach to linguistics influenced strongly NLP methods from the 1950s to the 1990s. Chomsky's approach to the "generative grammar" of language learning favored rules-based models of syntactic structure and, thus, supplied to computer science the grounds for devising algorithmic rules for the extraction of meaning from language. For an account of what was at stake in the 1950s debates between behavioral and rules-based approaches to language learning, see Chomsky, "Bad News," 26–58.

11　Amoore, *Politics of Possibility*, 39–45.

12　Statistical machine translation advanced significantly with the availability of labeled training data derived from the translations of all European Union documentation into the eleven recognized official languages of the EU. In effect, the human-translated documents provided a vast corpus of training data for translation algorithms. See Koehn, "Europarl." Advances in deep neural networks have developed these techniques beyond the statistical to devise predictive models for the recognition of human language. See Bahdanau, Cho, and Bengio, "Neural Machine Translation." For an overview of the contemporary state of the art in NLP methods, see Hirschburg and Manning, "Advances in Natural Language Processing," 261–66.

13 Kurzweil, *The Singularity Is Near.*

14 For the Google research team's account of the development of their algorithms see Pickett, Tar, and Strope, "On the Personalities of Dead Authors."

15 Juola, "How a Computer Program Helped Show."

16 Burgess, "Google Wants to Predict."

17 Amoore, "Data Derivatives," 24–43.

18 Cuthbertson, "Google's AI Predicts."

19 Foucault, "What Is an Author?," 378.

20 Foucault, "What Is an Author?," 378.

21 Foucault, "What Is an Author?," 378.

22 Foucault, "What Is an Author?," 379.

23 Foucault, "What Is an Author?," 377.

24 Foucault, "What Is an Author?," 380.

25 Foucault, "What Is an Author?," 381.

26 Foucault, "What Is an Author?," 383–85.

27 Foucault, "What Is an Author?," 383.

28 Foucault, "What Is an Author?," 383; my emphasis.

29 In Vicki Kirby's wonderful account of the potential for extension of deconstruction into the sciences of mathematics, physics, and biology, she argues that "deconstruction's home is as uncomfortable yet essential to the sciences as it is to the humanities." Kirby, *Quantum Anthropologies*, 6.

30 Foucault, "What Is an Author?," 380.

31 Foucault, "What Is an Author?," 385–86.

32 Turkle, *Second Self.*

33 Guizzo, "Google Wants Robots to Acquire."

34 Among the most troubling of the political consequences of the search for source code, private technology corporations have become very successful at protecting their algorithms as proprietary sources. What this means is that, even when our everyday interactions with algorithms generate their emergent capabilities, the imagination of a proprietary "source code" locks out the public from understanding this process.

35 Kirchner, "Federal Judge Unseals"; and Kirchner, "Traces of Crime." Judge Caproni had initially ruled that the vendor of the probabilistic genotyping technology had legitimate proprietary and copyright interests preventing the release of the source code. The ruling was subsequently overturned following a challenge from Yale Law School. The full documents are at "Memorandum in Support of Application."

36 Pishko, "Impenetrable Program."

37 For forensic protocols for the use of the FST algorithm, see "Forensic Biology Protocols."

38 Taylor, Bright, and Buckleton, "Source of Error," 33.

39 Adams, "Exhibit A."

40 Adams, "Exhibit A," 9.

41 Fowles, "Nature of Nature," 357.

42 Fowles, "Golding and 'Golding,'" 203. Edward Said similarly notes the character of great works of music and literature that cut against the grain of society's prevailing styles and sensibilities. Said, *On Late Style*.

43 Fowles, "Nature of Nature," 358.

44 Fowles, "Nature of Nature," 359.

45 Deleuze, *Negotiations*, 133.

46 Fowles, "Notes on an Unfinished Novel," 18.

47 "Writing unfolds like a game [*jeu*] that invariably goes beyond its own rules and transgresses its limits," writes Foucault. "In writing, the point is not to manifest or exalt the act of writing, nor is it to pin a subject within language; it is, rather, a question of creating a space into which the writing subject constantly disappears." Foucault, "What Is an Author?," 378.

48 Gilles Deleuze considers there to be "no literature without fabulation," so that "writing is a question of becoming, always incomplete, always in the midst of being formed." Deleuze, *Essays Critical and Clinical*, 1–3.

49 Critchley, *Ethics of Deconstruction*, 34.

50 Fowles and Vipond, "Unholy Inquisition," 368, 372.

51 Fowles and Vipond, "Unholy Inquisition," 372.

52 Critchley, *Ethics of Deconstruction*, 48.

53 Critchley, *Ethics of Deconstruction*, 48.

54 Fowles and Vipond, "Unholy Inquisition," 383.

55 Giorgio Agamben understands the author "as gesture" in the sense that "the author is present in the text only as a gesture that makes expression possible precisely by establishing a central emptiness within this expression." Agamben, "Author as Gesture," 66.

56 Mantel, "Day Is for the Living."

57 The concept of fabulation serves to further displace the figure of the author in favor of a focus on the power of writing. As Deleuze puts the problem: "There is no literature without fabulation, but as Bergson was able to see, fabulation—the fabulatory function—does not consist in imagining or projecting an ego. Rather, it attains these visions, it raises itself to these becomings and powers." Deleuze, *Essays Critical and Clinical*, 3.

58 Deleuze, *Essays Critical and Clinical*, 118.

59 Foucault, "Self Writing," 212.

60 Deleuze, *Negotiations*, 126.

61 The methods of deconstruction have long supplied to philosophy a means of engaging with an extended sense of writing, or what Jacques Derrida has called "archi-writing" (*archi-écriture*). Derrida, *Of Grammatology*, 15–20.

62 Derrida, *Limited Inc.*, 148.

63 Derrida, *Limited Inc.*, 19.

64 Derrida, *Limited Inc.*, 20.

65 Derrida, *Limited Inc.*, 152.

66 Hinton et al., "Deep Neural Networks," 83.

67 Hinton et al., "Deep Neural Networks," 84.

68 Critchley, *Ethics of Deconstruction*, 21.

69 Nissenbaum, *Privacy in Context.*

4. The Madness of Algorithms

1 On the notion of the flash crash and the advent of high-frequency trading algorithms, see MacKenzie, "How to Make Money in Microseconds," 16–18; Borch and Lange, "High Frequency Trader Subjectivity," 283–306; and Lange, Lenglet, and Seyfert, "Cultures of High-Frequency Trading," 149–65.

2 Gibbs, "Microsoft's Racist Chatbot."

3 Bridle, "Something Is Wrong on the Internet"; Lewis, "Fiction Is Outperforming Reality."

4 O'Neil, *Weapons of Math Destruction*; Parkin, "Teaching Robots Right from Wrong"; and Stubb, "Why Democracies Should Care."

5 Among the signatories of the Future of Life Institute's statement on artificial intelligence, Elon Musk and Ray Kurzweil adjudicate systems that "must do what we want them to do." Ormandy, "Google and Elon Musk to Decide."

6 When machine learning algorithms begin to generate their own form of language, the tipping point of terminating the system is reached when the human experimenters can no longer understand what the machines are saying to one another. This marking of a limit point at the threshold of human comprehension invokes a departure from what is considered reasonable, as seen, for example, when algorithms communicate with other algorithms. Walker, "Researchers Shut Down AI."

7 Derrida, "Cogito and the History of Madness," 46.

8 Derrida, "Cogito and the History of Madness," 39; and Foucault, *History of Madness.*

9 The recommendation system's architecture "funnels" candidate videos retrieved and ranked from a corpus of millions of videos, presenting a targeted set of recommendations to the user. Covington, Adams, and Sargin, "Deep Neural Networks."

10 Jacques Derrida famously opens his lecture on Foucault's *History of Madness* with an epigraph he attributes to Kierkegaard: "The instant of the decision is madness." He uses different formulations of Kierkegaard's decision across a number of his works. For Derrida, all decision worthy of the name is a decision made without reference to programmed knowledge, in a state of nonknowledge that is darkness. Derrida's use of Kierkegaard does not quite accurately capture the paradox of decision as Kierkegaard formulates it. Derrida, "Cogito and the History of Madness," 36; and Bennington, "Moment of Madness," 103–27.

11 Thomas Keenan addresses directly the politics of decision as an "ethics and poli-

tics" without grounds, so that "we have politics because we have no grounds, no reliable standpoints." Keenan, *Fables of Responsibility*, 4.

12 Alpaydin, *Machine Learning*, 79.

13 Norbert Wiener, "Atomic Knowledge of Good and Evil" [1950], folder 631, box 28D, MC 22, Norbert Wiener Papers, Institute Archives and Special Collections, MIT Libraries, Cambridge, Massachusetts (hereafter Wiener Papers).

14 Kate Crawford, "Dark Days"; Wiener, "Scientist Rebels"; Norbert Wiener, "Rebellious Scientist after Two Years," *Bulletin of the Atomic Sciences*, folder 601, box 28C, Wiener Papers.

15 Norbert Wiener, "The Human Use of Human Beings" [1950], folder 639, box 29A, MC 22, Wiener Papers.

16 Erickson et al., *How Reason Almost Lost Its Mind*, 2.

17 Erickson et al., *How Reason Almost Lost Its Mind*, 30.

18 Norbert Wiener, "Second Industrial Revolution and the New Concept of the Machine," folder 619, box 28D, MC 22, Wiener Papers.

19 O. Halpern, *Beautiful Data*, 150.

20 O. Halpern, *Beautiful Data*, 151.

21 Chun, *Control and Freedom*, 300; Liu, *Freudian Robot*, 202.

22 M. Beatrice Fazi argues that computation is a process of determining indeterminacy, where the generative potential of algorithms works with Turing's notion of the incomputable. Fazi, *Contingent Computation*. The kind of generative potential identified by Fazi also drives the capacity of algorithmic systems to derive value from speculative and experimental models.

23 Foucault, *History of Madness*, 164.

24 Foucault, *History of Madness*, xxxii.

25 Foucault, *History of Madness*, 169.

26 Jacques Derrida describes Foucault's determination to avoid the trap of a history of madness untamed by reason as "the maddest aspect of his project." Derrida, "Cogito and the History of Madness," 45; and Foucault, "Reply to Derrida," 575.

27 Derrida, "Cogito and the History of Madness," 46.

28 Derrida, "Cogito and the History of Madness," 46.

29 "To the extent that madness is a form of knowledge," writes Foucault, "it was also a set of norms, both norms against which madness could be picked out as a phenomenon of deviance within society, and, at the same time, norms of behaviour for normal individuals." The idea of madness, for Foucault, defined "a certain mode of being of the normal subject, as opposed to and in relation to the mad subject." The dissension between madness and reason, then, also marks the relation of selves to others in relation to thresholds of norm and anomaly. Foucault, *Government of Self and Others*, 3.

30 Foucault, *History of Madness*, 426.

31 Foucault, *History of Madness*, 426.

32 Foucault, *History of Madness*, 461.

33 Foucault, *History of Madness*, 159.

34 Foucault, *History of Madness*, 527.

35 Foucault, *History of Madness*, 528.

36 Foucault, *History of Madness*, 164.

37 Derrida, "Cogito and the History of Madness," 62–63.

38 In place of the notion that the madness of algorithms denotes a departure from some normal course, I envisage something closer to Brian Rotman's concept of "becoming beside oneself." Rotman's concept depicts a plurality of being in which the self is "distributed and in excess of unity." Rotman, *Becoming Beside Ourselves*, 103.

39 The NSF funded Nicholas Evans's study "Ethical Algorithms in Autonomous Vehicles" in 2017. See NSF Award Abstract 1734521. See also Nyholm and Snids, "Ethics of Accident-Algorithms," 1275–89.

40 Phys.org, "UMass Lowell Professor Steers Ethical Debate." The encoded ethics of algorithms for autonomous vehicles poses questions such as "Should your self-driving car protect you at all costs? Or should it steer you into a ditch to avoid hitting a school bus full of children?" The notion of a moral choice to be made and to be encoded within the algorithm looms large in these accounts of the ethics of decision. Webster, "Philosophy Professor Wins NSF Grant."

41 Phys.org, "UMass Lowell Professor Steers Ethical Debate."

42 The school bus features discursively within many experimental scenarios for the development of ethical codes governing algorithmic action. The image of a school bus embodies the fears of algorithms recognizing objects for drone surveillance, and the school bus similarly serves as the epitome of a mad decision within autonomous vehicles research.

43 Derrida, *Gift of Death*, 24.

44 Eklund, Nichols, and Knutson, "Cluster Failure," 7900–7905.

45 Timmer, "Software Faults Raise Questions."

46 Brackeen, "Facial Recognition Software."

47 Foucault, *History of Madness*, 241.

48 Domingos, *Master Algorithm*, 83.

49 Kourou et al., "Machine Learning Applications," 8–17.

50 As both Derrida and Foucault elaborate, madness is what cannot be said, the unspoken that is absent from a body of work. Derrida, "Cogito and the History of Madness," 51.

51 In all the domains where I have followed the development of algorithms for state security—border controls, counter-terror, policing, and immigration control—my discussions with algorithm designers have included some reference to the use of random forest algorithms as preferred by operational security teams. Though I cannot definitively explain the allure these algorithms hold for the state, in many cases, the random forest algorithm was using proxy data to identify patterns associated with potential threats.

52 Grothoff and Porup, "NSA's SKYNET Program."

53 Grothoff and Porup, "NSA's SKYNET Program," 2.

54 Amoore, "Data Derivatives," 24–43.

55 Breiman, "Random Forests," 5–32. Breiman's paper has had extraordinary in-

fluence on the development of machine learning, having been cited more than thirty-five thousand times.

56 Breiman, "Random Forests," 6.

57 Breiman, "Random Forests," 7. See also Abdulsalam, Skillicorn, and Martin, "Classification Using Streaming Random Forests," 24.

58 Daston, "Probabilistic Expectation," 234–60.

59 Grothoff and Porup, "NSA's SKYNET Program," 3.

60 Abdulsalam, Skillicorn, and Martin, "Classification Using Streaming Random Forests," 25.

61 Tucker, "New AI Learns through Observation Alone."

62 Foucault, *History of Madness*, 521.

63 It has long been acknowledged in relation to financial markets that notions of irrationality or chance have a place within the making of ideas of rational economic humans. De Goede, *Virtue, Fortune, and Faith*. Derrida, *Death Penalty*, 21.

64 Foucault, *History of Madness*, 521.

65 For Derrida, "calculation is always calculating with what is incalculable as well as what is non-calculable" so that there is always also indecision and undecidability. Derrida, *Death Penalty*, 197.

66 Derrida, *Death Penalty*, 4, 42.

67 Derrida, *Death Penalty*, 93.

68 Derrida's discussion of the scaffold invokes the material scaffold of the sovereign scene of the death penalty as well as the "scaffolding" that renders an act as an unconditional principle. Derrida, *Death Penalty*, 21.

69 Derrida, *Death Penalty*, 26.

70 Derrida, *Death Penalty*, 6, 139.

5. The Doubtful Algorithm

1 While working on the Manhattan Project, Feynman and his colleagues calculated the energy release from the atomic bomb's implosion using IBM punch card machines. Feynman describes how they learned to compute "around the error" when the "machine made a mistake." The use of such calculating machines allowed the scientists to conduct multiple calculations in parallel so that "a mistake made at some point in the cycle only affects the nearby numbers." Whereas in their early attempts at the mathematics of the atomic bomb, the linearity of the calculation meant that errors caused them to "go back and do it over again," the parallel processing of the punch cards allowed them to incorporate the error and continue to progress with the program. In short, intrinsic to the science of the atomic bomb was a form of computation that could incorporate error and continue to a useable output. Some years later, after the events of Hiroshima and Nagasaki, Feynman reflected on the relationship between his partial and experimental computation and the science of the atomic bomb: "I looked out at the buildings and I began to think, you know, about how much the radius of the Hiroshima bomb damage was. . . . How far from here was 34th Street? It's a terrible thing that we made." Feynman, *Surely You're Joking*, 131–36.

2 Feynman, *What Do You Care What Other People Think?*, 245, 247.

3 Braidotti, "Posthuman, All Too Human," 197.

4 Haraway, *Modest_Witness@Second_Millennium*, 7.

5 I owe the development of this point on the enactment of a decision indifferent to persistent doubt to discussions with Susan Schuppli, Eyal Weizman, and their wonderful postgraduate forensic architecture students, February 19, 2016.

6 As Rosi Braidotti argues, the posthuman subject need not be anthropocentric, but "it must be the site for political and ethical accountability." Braidotti, *Posthuman*, 103.

7 Haraway, *Modest_Witness@Second_Millennium*, 23–24.

8 Butler, *Giving an Account of Oneself*, 29.

9 House of Commons Science and Technology Select Committee, *Algorithms in Decision Making*.

10 Fast learning algorithms used in systems such as facial recognition are increasingly generating their own "ground truths" through their exposure to data. Among the many computer science examples of this process, see Yang et al., "Fast ℓ_1-Minimization Algorithms," 3234–46; and Hinton, Osindero, and Teh, "Fast Learning Algorithm," 1527–54.

11 Alpaydin, *Machine Learning*, 116.

12 Bottou, "From Machine Learning to Machine Reasoning," 133–49.

13 Phil Garnett, of York University UK, put this point to me in response to my account of how an image recognition neural network learns to recognize form. He captures here the significance of the algorithm's indifference to the attributes of the cat as qualities or essence. The relation to the ground truth data means that the algorithm does not need to have a truth of the essence of "catness," just as it is indifferent to the truthfulness of a person who is subject to an algorithmic decision about their propensity for crime or the repaying of debts.

14 B. Anderson, "Preemption, Precaution, Preparedness"; Aradau and Van Munster, *Politics of Catastrophe*; de Goede, *Speculative Security*; and Amoore, *Politics of Possibility*.

15 Haraway, "Situated Knowledges," 581; and Hayles, *How We Became Posthuman*, 287.

16 Interview with designers of deep neural networks, London, September 2014.

17 The target output of the algorithm is, for this reason, akin to Samuel Weber's concept of a "target of opportunity," in which the singularity of the targeted strike is defined in terms of the realization of opportunities. With regard to algorithms, the target of opportunity involves an extractive economy of profit and sovereign power from the target output. S. Weber, *Targets of Opportunity*, 5.

18 Woodman, "Palantir Provides the Engine."

19 House of Representatives Committee on Science and Technology, *Investigation of the Challenger Accident*.

20 Feynman, *What Do You Care What Other People Think?*, 113.

21 Erickson et al., *How Reason Almost Lost Its Mind*, 29–30.

22 Feynman, *What Do You Care What Other People Think?*, 17.

23 Vaughan, *Challenger Launch Decision*, xiv–xv.

24 P. Halpern, *Quantum Labyrinth*, 114.

25 Barad, "On Touching," 207.

26 Feynman, *What Do You Care What Other People Think?*, 135.

27 Feynman, *What Do You Care What Other People Think?*, 137.

28 Feynman, *What Do You Care What Other People Think?*, 107.

29 Haraway, "Situated Knowledges," 584.

30 Feynman, *What Do You Care What Other People Think?*, 180.

31 Bennett, *Vibrant Matter*.

32 Heidegger, "Bauen, Wohnen, Denken."

33 Hayles, *Unthought*, 29.

34 Latour and Weibel, *Making Things Public*.

35 Haraway, *Modest_Witness@Second_Millennium*, 24, 34.

36 Barad, "On Touching," 208.

37 Haraway, *Modest_Witness@Second_Millennium*, 35.

38 Foucault, *Courage of Truth*, 79.

39 Foucault, *Courage of Truth*, 30.

40 Butler, *Giving an Account of Oneself*, 19.

41 Foucault, *Government of Self and Others*, 62.

42 Foucault, *Wrong-Doing, Truth-Telling*, 200.

43 Feynman, *What Do You Care What Other People Think?*, 139.

44 Berlant, *Cruel Optimism*, 23–24.

45 Berlant, *Cruel Optimism*, 262.

46 There is widespread acknowledgement among self-styled data scientists that 90 percent of the work of building a model remains a task of data preparation.

47 Parisi, *Contagious Architecture*, 95.

48 Feynman, *What Do You Care What Other People Think?*, 214.

49 Harding, *Snowden Files*; and Greenwald, *No Place to Hide*.

50 D. Anderson, *Attacks in London and Manchester*.

51 Keenan, *Fables of Responsibility*, 3, 42.

52 Keenan, *Fables of Responsibility*, 12; and Derrida, "Force of Law," 26.

53 Observation of presentations of "threat intelligence" software to UK government, London, June 2017.

54 Truvé, "Building Threat Analyst Centaurs."

55 Butler, *Giving an Account of Oneself*, 19, 45.

56 Derrida, *Gift of Death*, 24; and Keenan, *Fables of Responsibility*, 2.

57 Deleuze, *Bergsonism*, 29.

58 Barad, "On Touching," 212.

59 Barad, *Meeting the Universe Halfway*, 265.

60 Feynman, *Surely You're Joking*, 247; my emphasis.

6. The Unattributable

1 Xu et al., "Show, Attend and Tell."

2 Vinyals et al., "Show and Tell," 2048–57.

3 Xu et al., "Show, Attend and Tell," 2048.

4 Vinyals et al., "Show and Tell," 653.

5 Crary, *Suspensions of Perception.*

6 Xu et al., "Show, Attend and Tell," 2048.

7 Vinyals et al., "Show and Tell," 652.

8 Xu et al., "Show, Attend and Tell," 2048.

9 Derrida, *Limited Inc.,* 152–53.

10 Shane and Wakabayashi, "Business of War."

11 Derrida, *Limited Inc.,* 151.

12 Deleuze, *Essays Critical and Clinical,* 4.

13 Deleuze, *Essays Critical and Clinical,* 113.

14 Bottou, "From Machine Learning to Machine Reasoning," 133–49.

15 To begin from the propensities of the algorithm, and not from strategies to overturn it, is to echo Isabelle Stengers's sense that "the diagnosis of becomings" is not a strategy for alternatives but rather a "speculative operation, thought experiment" in which the possibilities are not determined. Stengers, *Cosmopolitics I,* 12–13.

16 Addressing the machine learner "kittydar," Adrian Mackenzie describes how "the software finds cats by cutting the images into smaller windows. For each window, it measures a set of gradients" and then "compares these measurements to the gradients of known cat images." Mackenzie, *Machine Learners,* 4–5.

17 The UK Information Commissioner's Office (ICO) is investigating the use of millions of Facebook profiles to train the Cambridge Analytica algorithms. At the time of writing, the maximum fine has been imposed on Facebook for the unlawful sharing of data. See Greenfield, "Cambridge Analytica Files."

18 Deleuze, *Bergsonism,* 16.

19 The question, as Henri Bergson formulated it, "is not how perception arises, but how is it limited," so that the image of a whole scene is "reduced to the image of that which interests you." Bergson, *Matter and Memory,* 34.

20 For Crary, perception exceeds the "single-sense modality of sight," extending to "hearing and touch" and "most importantly, irreducibly mixed modalities." Crary, *Suspensions of Perception,* 1, 4.

21 Deleuze, *Bergsonism,* 31.

22 Deleuze, *Bergsonism,* 32.

23 Simon Critchley argues that "deconstruction is a philosophy of hesitation" in that it carries with it "the experience of undecidability" essential to ethics. Critchley, *Ethics of Deconstruction,* 42.

24 For Derrida, there is a close relation between "feeling the weight" of something, "peser," and thought, "penser." Derrida, *Death Penalty,* 93.

25 For Derrida, the "trace of the rejected alternative" is not a matter of "either this or that" in the sense that some alternative path could be reopened, but a "neither this . . . nor that," where the decisive moment is unknowable. Derrida, *Writing and Difference,* 112.

26 Keenan, *Fables of Responsibility,* 1, 51.

27 Keenan, *Fables of Responsibility,* 51.

28 Writing on the genealogy of ethics, Michel Foucault describes the transformation in how Greek and Christian societies viewed themselves: these changes "are not in the code, but in what I call the 'ethics'" which is in "their relations to themselves and to others." Foucault, "On the Genealogy of Ethics," 255, 266.

29 Deleuze, *Spinoza*, 125.

30 Deleuze, *Spinoza*, 125.

31 Stengers, *Thinking with Whitehead*, 515.

32 Critchley, *Ethics of Deconstruction*, 18.

33 Derrida, *Limited Inc.*, 122.

34 Haraway, "Situated Knowledges," 575–99.

35 Haraway, "Situated Knowledges," 586.

36 Haraway, "Situated Knowledges," 583.

37 Haraway, "Situated Knowledges," 586.

38 Xu et al., "Show, Attend and Tell," 2052, 2055.

39 Burgess, "UK Police Are Using AI."

40 Laughland, "Baltimore Unrest."

41 Derrida, *Limited Inc.*, 152–53.

42 Deleuze, *Spinoza*, 17, 51.

43 Foucault, "What Is Critique?," 265, 278.

44 Critchley, *Ethics of Deconstruction*, 48.

45 Critchley, *Ethics of Deconstruction*, 48.

46 As Deleuze has written, the singularities "connect with one another in a manner entirely different from how individuals connect" and can be neither clustered nor "grouped [or] divided in the same way." Deleuze, *Pure Immanence*, 30.

Abdulsalam, Hanady, David B. Skillicorn, and Patrick Martin. "Classification Using Streaming Random Forests." *IEEE Transactions on Knowledge and Data Engineering* 23, no. 1 (2011): 24.

"About." PCL. Accessed April 2018. http://www.pointclouds.org/about/.

ACLU. "Baltimore County Police Department and Geofeedia Partner to Protect the Public during Freddie Gray Riots." ACLU, October 11, 2016. https://www.aclunc .org/docs/20161011_geofeedia_baltimore_case_study.pdf.

Adams, Nathaniel. "Exhibit A: Declaration of Nathaniel Adams." October 17, 2017. https://www.documentcloud.org/documents/4112649-10-17-17-Unredacted-NA -Exhibit-A.html.

Adey, Peter. *Aerial Life: Spaces, Mobilities, Affects.* Oxford: Wiley-Blackwell, 2010.

Agamben, Giorgio. "The Author as Gesture." In *Profanations,* translated by Jeff Fort, 61–72. New York: Zone Books, 2007.

Agamben, Giorgio. *Potentialities: Collected Essays in Philosophy.* Stanford, CA: Stanford University Press, 2007.

Allen, John. "The Whereabouts of Power: Politics, Government and Space." *Geografiska Annaler: Series B* 86, no. 1 (2004): 19–32.

Alpaydin, Ethem. *Machine Learning.* Cambridge, MA: MIT Press, 2016.

Alpers, Svetlana. "The Studio, the Laboratory and the Vexations of Art." In *Picturing Science, Producing Art,* edited by P. Galison and C. Jones, 401–17. New Haven, CT: Yale University Press, 1998.

Amaro, Ramon. "Pre-emptive Citizenship: Crime Prediction and Statistical Intervention in Racialized Ecologies." Paper presented at the Radical Ecologies Conference, Goldsmiths University of London, May 4, 2015.

Amaro, Ramon. "Race, Surveillance and Autopoetically Instituted Being." Paper presented at the Race and Surveillance Conference, Birkbeck University of London, May 4, 2018.

Amoore, Louise. "Data Derivatives: On the Emergence of a Security Risk Calculus for Our Times." *Theory, Culture and Society* 28, no. 6 (2011): 24–43.

Amoore, Louise. *The Politics of Possibility: Risk and Security beyond Probability.* Durham, NC: Duke University Press, 2013.

Amoore, Louise, and Marieke de Goede. "Transactions after 9/11: The Banal Face of the Preemptive Strike." *Transactions of the Institute of British Geographers* 33, no. 2 (2008): 173–85.

Amoore, Louise, and Volha Piotukh. "Life beyond Big Data: Governing with Little Analytics." *Economy and Society* 44, no. 3 (2015): 341–66.

Anderson, Ben. "Preemption, Precaution, Preparedness: Anticipatory Action and Future Geographies." *Progress in Human Geography* 34, no. 6 (2010): 777–98.

Anderson, C. "The End of Theory: The Data Deluge Makes Scientific Method Obsolete." *Wired,* June 23, 2008.

Anderson, David. *Attacks in London and Manchester.* London: Brick Court Chambers, 2017. https://assets.publishing.service.gov.uk/government/uploads/system/uploads/attachment_data/file/664682/Attacks_in_London_and_Manchester_Open_Report.pdf.

Angwin, Julia, Jeff Larson, Surya Mattu, and Lauren Kirchner. "Machine Bias." *ProPublica*, May 23, 2016. https://www.propublica.org/article/machine-bias-risk-assessments-in-criminal-sentencing.

Aradau, Claudia, and Rens Van Munster. *Politics of Catastrophe: Genealogies of the Unknown.* London: Routledge, 2011.

Ash, James. *The Interface Envelope: Gaming, Technology, Power.* London: Bloomsbury, 2015.

Bahdanau, Dzmitry, Kyunghyun Cho, and Yoshua Bengio. "Neural Machine Translation by Jointly Learning to Align and Translate." Presented at the International Conference on Learning Representations, San Diego, California, May 7–9, 2015.

Bal, Mieke. "The Commitment to Look." *Journal of Visual Culture* 4, no. 2 (2005): 145–62.

Barad, Karen. *Meeting the Universe Halfway: Quantum Physics and the Entanglement of Matter and Meaning.* Durham, NC: Duke University Press, 2007.

Barad, Karen. "On Touching: The Inhuman That Therefore I Am." *Differences: A Journal of Feminist Cultural Studies* 23, no. 3 (2012): 206–23.

Beer, David. "Power through the Algorithm? Participatory Web Cultures and the Technological Unconscious." *New Media and Society* 11, no. 6 (2009): 985–1002.

Bennett, Jane. *Vibrant Matter.* Durham, NC: Duke University Press, 2010.

Bennington, Geoffrey. "A Moment of Madness: Derrida's Kierkegaard." *Oxford Literary Review* 3, no. 1 (2011): 103–27.

Bergson, Henri. *The Creative Mind: An Introduction to Metaphysics.* 1946. New York: Dover, 2007.

Bergson, Henri. *Matter and Memory.* 1912. Mineola, NY: Dover Philosophical Classics, 2004.

Berlant, Lauren. *Cruel Optimism.* Durham, NC: Duke University Press, 2011.

Berlinski, David. *The Advent of the Algorithm.* New York: Harcourt, 2000.

Berry, David. *The Philosophy of Software.* Basingstoke: Macmillan, 2011.

Borch, Christian, and Ann-Christina Lange. "High Frequency Trader Subjectivity: Emotional Attachment and Discipline in an Era of Algorithms." *Socio-Economic Review* 15 (2017): 283–306.

Bottou, L. "From Machine Learning to Machine Reasoning." *Machine Learning* 94 (2014): 133–49.

boyd, danah, and Kate Crawford. "Critical Questions for Big Data." *Information, Communication and Society* 15, no. 5 (2012): 662–79.

Brackeen, Brian. "Facial Recognition Software Is Not Ready for Use by Law Enforcement." *TechCrunch*, June 22, 2018.

Braidotti, Rosi. *The Posthuman.* Cambridge: Polity, 2013.

Braidotti, Rosi. "Posthuman, All Too Human: Towards a New Process Ontology." *Theory, Culture and Society* 23, no. 7–8 (2006): 197–208.

Brandom, Russell. "A New System Can Measure the Hidden Bias in Otherwise Secret Algorithms." *Verge*, May 25, 2016. https://www.theverge.com/2016/5/25/11773108/research-method-measure-algorithm-bias.

Bratton, Benjamin. *The Stack: On Software and Sovereignty.* Cambridge, MA: MIT Press, 2015.

Breiman, Leo. "Random Forests." *Machine Learning* 45, no. 1 (2001): 5–32.

Bridge, Mark, and Gabriella Swerling. "Bail or Jail? App Helps Police Make Decision about Suspect." *Sunday Times* (London), May 11, 2017. https://www.thetimes.co.uk/article/bail-or-jail-app-helps-police-make-decision-about-suspect-kv766zjc9.

Bridle, James. *New Dark Age: Technology and the End of the Future.* London: Verso, 2018.

Bridle, James. "Something Is Wrong on the Internet." *Medium*, November 6, 2017.

Browne, Simone. *Dark Matters: On the Surveillance of Blackness.* Durham, NC: Duke University Press, 2015.

Bucher, Taina. *If . . . Then: Algorithmic Power and Politics.* Oxford: Oxford University Press, 2018.

Burgess, Matt. "Google Wants to Predict the Next Sentences of Dead Authors." *Wired* (UK), February 26, 2016. https://www.wired.co.uk/article/google-author-create-artificial-intelligence.

Burgess, Matt. "UK Police Are Using AI to Inform Custodial Decisions—But It Could Be Discriminating against the Poor." *Wired*, March 1, 2018.

Butler, Judith. *Giving an Account of Oneself: A Critique of Ethical Violence.* Amsterdam: Van Gorcum Press, 2003.

Butler, Judith. *Precarious Life: The Powers of Mourning and Violence.* London: Verso, 2004.

Caruso, Catherine. "Can a Social Media Algorithm Predict a Terror Attack?" *MIT Technology Review*, June 16, 2016. https://www.technologyreview.com/s/601700/can-a-social-media-algorithm-predict-a-terror-attack/.

Central Intelligence Agency. "CIA Creates a Cloud: An Interview with CIA's Chief Intelligence Officer, Doug Wolfe, on Cloud Computing at the Agency." CIA Featured Story Archive, 2014. https://www.cia.gov/news-information/featured-story-archive/2014-featured-story-archive/cia-creates-a-cloud.html.

Cheney-Lippold, John. *We Are Data: Algorithms and the Making of Our Digital Selves.* New York: New York University Press, 2019.

Chomsky, Noam. "Bad News: Noam Chomsky—A Review of B. F. Skinner's Verbal Behavior." *Language* 35, no. 1 (1959): 26–58.

Chun, Wendy Hui Kyong. *Control and Freedom: Power and Paranoia in the Age of Fiber Optics.* Cambridge, MA: MIT Press, 2006.

Comaniciu, Dorin, and Peter Meer. "Mean Shift: A Robust Approach toward Feature Space Analysis." *IEEE Transaction on Pattern Analysis and Machine Intelligence* 24, no. 5 (2002): 605.

Connolly, William. "Beyond Good and Evil: The Ethical Sensibility of Michel Foucault." *Political Theory* 21, no. 3 (1993): 365–89.

Covington, Paul, Jay Adams, and Emre Sargin. "Deep Neural Networks for YouTube Recommendations." Paper presented at the 10th ACM Conference on Recommender Systems, Boston, Massachusetts, September 15–19, 2016.

Cowen, Deborah. *The Deadly Life of Logistics.* Minneapolis: University of Minnesota Press, 2014.

Crampton, Jeremy. "Cartographic Calculations of Territory." *Progress in Human Geography* 35, no. 1 (2010): 92–103.

Crary, Jonathan. *24/7: Late Capitalism and the End of Sleep.* London: Verso, 2013.

Crary, Jonathan. *Suspensions of Perception: Attention, Spectacle, and Modern Culture.* Cambridge, MA: MIT Press, 2001.

Crawford, Kate. "Artificial Intelligence's White Guy Problem." *New York Times*, June 25, 2016.

Crawford, Kate. "Dark Days: Artificial Intelligence and the Rise of Fascism." Presented at the SXSW Conference, Austin, Texas, March 12, 2017.

Crawford, Kate, and Jason Schultz. "Big Data and Data Protection: Toward a Framework to Redress Predictive Privacy Harms." *Boston College Law Review* 55, no. 1 (2014): 93–128.

Critchley, Simon. *The Ethics of Deconstruction: Derrida and Levinas.* Oxford: Blackwell, 1992.

Cuthbertson, Anthony. "Google's AI Predicts the Next Sentence of Dead Authors." *Newsweek*, February 29, 2016.

Daston, Lorraine. "The Moral Economy of Science." *Osiris* 10 (1995): 2–24.

Daston, Lorraine. "Probabilistic Expectation and Rationality in Classical Probability Theory." *Historia Mathematica* 7 (1980): 234–60.

Daston, Lorraine, and Peter Galison. *Objectivity.* Cambridge, MA: MIT Press, 2007.

de Goede, Marieke. *Speculative Security: The Politics of Pursuing Terrorist Monies.* Minneapolis: University of Minnesota Press, 2012.

de Goede, Marieke. *Virtue, Fortune, and Faith: A Genealogy of Finance.* Minneapolis: University of Minnesota Press, 2005.

de Goede, Marieke, and Samuel Randalls. "Precaution, Preemption: Arts and Technologies of the Actionable Future." *Environment and Planning D: Society and Space* 27, no. 5 (2009): 859–78.

DeLanda, Manuel. *Intensive Science and Virtual Philosophy.* London: Continuum, 2002.

DeLanda, Manuel. *Philosophy and Simulation.* London: Bloomsbury, 2011.

Deleuze, Gilles. *Bergsonism.* New York: Zone Books, 1991.

Deleuze, Gilles. *Essays Critical and Clinical.* Translated by Ariel Greco and Daniel Smith. Minneapolis: University of Minnesota Press, 1998.

Deleuze, Gilles. *Negotiations: 1972–1990.* Translated by Martin Joughin. New York: Columbia University Press, 1995.

Deleuze, Gilles. "On the Difference between the *Ethics* and a Morality." In *Spinoza: Practical Philosophy*, translated by Robert Hurley, 17–28. San Francisco, CA: City Lights, 1988.

Deleuze, Gilles. *Pure Immanence: Essays on a Life.* New York: Zone Books, 2001.

Deleuze, Gilles. *Spinoza: Practical Philosophy.* Translated by Robert Hurley. San Francisco, CA: City Lights, 1988.

Derrida, Jacques. *Archive Fever: A Freudian Impression.* Chicago: University of Chicago Press, 1995.

Derrida, Jacques. "Cogito and the History of Madness." In *Writing and Difference*, translated by A. Bass, 31–63. London: Routledge, 2001.

Derrida, Jacques. *The Death Penalty.* Vol. 2. Chicago: University of Chicago Press, 2016.

Derrida, Jacques. "Force of Law: The Mystical Foundations of Authority." In *Deconstruction and the Possibility of Justice*, edited by D. Cornell, M. Rosenfeld, and D. G. Carlson, 26. London: Routledge, 1992.

Derrida, Jacques. *The Gift of Death.* Chicago: University of Chicago Press, 1996.

Derrida, Jacques. *Limited Inc.* Evanston, IL: Northwestern University Press, 1988.

Derrida, Jacques. *Of Grammatology.* Translated by Gayatri Spivak. Baltimore, MD: Johns Hopkins University Press, 1976.

Derrida, Jacques. *Writing and Difference.* London: Routledge, 2001.

Devlin, Keith. *The Math Gene.* London: Weidenfeld and Nicolson, 2000.

"DHS FOIA Documents Baltimore Protests Freddie Gray." Scribd. Uploaded by Jason Leopold. Accessed July 10, 2017. https://www.scribd.com/document/274209838/DHS-FOIA-Documents-Baltimore-Protests-Freddie-Gray.

Diamond, Cora, ed. *Wittgenstein's Lectures on the Foundations of Mathematics, Cambridge, 1939.* Chicago: University of Chicago Press, 1976.

Digital Reasoning. *Synthesys Mission Analytics.* Franklin, TN: Digital Reasoning, 2015.

Dodge, Martin. "Seeing Inside the Cloud: Some Ways to Map the Internet." Media Art Net Lectures, Karlsruhe, January 24, 2004.

Dodge, Martin, and Rob Kitchin. *Mapping Cyberspace.* London: Routledge, 2001.

Domingos, Pedro. *The Master Algorithm: How the Quest for the Ultimate Learning Machine Will Remake Our World.* New York: Allen Lane, 2015.

Dunning, Ted, and Ellen Friedman. *Practical Machine Learning: A New Look at Anomaly Detection.* Sebastopol, CA: O'Reilly, 2014.

Dürrenmatt, Friedrich. *The Physicists.* New York: Grove, 2006.

Eklund, Anders, Thomas Nichols, and Hans Knutson. "Cluster Failure: Why fMRI Inferences for Spatial Extent Have Inflated False-Positive Rates." *Proceedings of the National Academy of Sciences of the United States of America* 113, no. 28 (2016): 7900–7905.

Elden, Stuart. "Land, Terrain, Territory." *Progress in Human Geography* 34, no. 6 (2010): 799–817.

Elden, Stuart. *Mapping the Present: Heidegger, Foucault and the Project of a Spatial History.* London: Continuum, 2001.

Elden, Stuart. "Secure the Volume: Vertical Geopolitics and the Depths of Power." *Political Geography* 34, no. 1 (2013): 35–51.

Emerging Technology from the arXiv. "Twitter Data Mining Reveals the Origins of Support for Islamic State." *MIT Technology Review*, March 23, 2015. https://www.technologyreview.com/s/536061/twitter-data-mining-reveals-the-origins-of-support-for-islamic-state/.

Erickson, Paul J., Judy Klein, Lorraine Daston, Rebecca Lemov, Thomas Sturm, and Michael D. Gordin, eds. *How Reason Almost Lost Its Mind: The Strange Career of Cold War Reasoning.* Chicago: University of Chicago Press, 2013.

Eubanks, Virginia. *Automating Inequality: How High-Tech Tools Profile, Police, and Punish the Poor*. New York: Palgrave Macmillan, 2018.

European Commission. "What Does the Commission Mean by Secure Cloud Computing Services in Europe?" Memo 13/898, Brussels, October 15, 2013. http://europa.eu/rapid/press-release_MEMO-13-898_en.htm.

Fazi, M. Beatrice. *Contingent Computation: Abstraction, Experience, and Indeterminacy in Computational Aesthetics*. Lanham, MD: Rowman and Littlefield, 2018.

Feynman, Richard. *Surely You're Joking, Mr. Feynman!: Adventures of a Curious Character*. New York: Norton, 1985.

Feynman, Richard. *What Do You Care What Other People Think?: Further Adventures of a Curious Character*. New York: Norton, 1988.

Flinders, Karl. "Swift Builds Underground Datacentre to Keep Bank Transactions in Europe." *Computer Weekly*, April 30, 2012.

"Forensic Biology Protocols for Forensic STR Analysis." City of New York website, June 20, 2016. http://www1.nyc.gov/assets/ocme/downloads/pdf/technical-manuals/protocols-for-forensic-str-analysis/forensic-statistical-tool-fst.pdf.

Foucault, Michel. *The Courage of Truth: Lectures at the Collège de France, 1983-84*. Basingstoke: Palgrave, 2011.

Foucault, Michel. "The Ethics of the Concern for Self as a Practice of Freedom." In *Ethics: Essential Works of Foucault, 1954-1984*, edited by Paul Rabinow, 281-302. London: Penguin, 1997.

Foucault, Michel. *The Government of Self and Others: Lectures at the Collège de France, 1982-83*. Basingstoke: Palgrave, 2010.

Foucault, Michel. *History of Madness*. Edited by J. Khalfa. Translated by J. Murphy and J. Khalfa. London: Routledge, 2006.

Foucault, Michel. "On the Genealogy of Ethics: An Overview of Work in Progress." In *Ethics: Essential Works of Foucault 1954-1984*, edited by Paul Rabinow, 253-80. London: Penguin, 1997.

Foucault, Michel. "Polemics, Politics, Problematizations." In *Ethics: Essential Works of Foucault, 1954-1984*, edited by Paul Rabinow, 115. London: Penguin, 1997.

Foucault, Michel. "Reply to Derrida." In *History of Madness*, edited by J. Khalfa, translated by J. Murphy and J. Khalfa, 575. London: Routledge, 2006.

Foucault, Michel. *Security, Territory, Population: Lectures at the Collège de France, 1977-1978*. London: Palgrave Macmillan, 2004.

Foucault, Michel. "Self Writing." In *Ethics: Essential Works of Foucault, 1954-1984*, edited by P. Rabinow, 207-22. London: Penguin, 2000.

Foucault, Michel. "What Is an Author?" In *Aesthetics, Method, and Epistemology*, vol. 2 of *Essential Works of Foucault, 1954-1984*, edited by James Faubion, 205-22. New York: Penguin, 2000.

Foucault, Michel. "What Is Critique?" In *The Essential Foucault: Selections from the Essential Works of Foucault, 1954-1984*, edited by Paul Rabinow and Nikolas Rose, 263-78. New York: New Press, 2003.

Foucault, Michel. *Wrong-Doing, Truth-Telling: The Function of Avowal in Justice*. Edited by Fabienne Brion and Bernard Harcourt. Chicago: University of Chicago Press, 2014.

Fowles, John. "Golding and 'Golding.'" In *Wormholes: Essays and Occasional Writings*, edited by Jan Relf, 197–207. London: Jonathan Cape, 1998.

Fowles, John. "The Nature of Nature." In *Wormholes: Essays and Occasional Writings*, edited by Jan Relf, 343–62. London: Jonathan Cape, 1998.

Fowles, John. "Notes on an Unfinished Novel." In *Wormholes: Essays and Occasional Writings*, edited by Jan Relf, 13–26. London: Jonathan Cape, 1998.

Fowles, John, and Susanna Onega. *Form and Meaning in the Novels of John Fowles*. London: VMI Press, 1989.

Fowles, John, and Dianne Vipond. "An Unholy Inquisition." In *Wormholes: Essays and Occasional Writings*, edited by Jan Relf, 365–84. London: Jonathan Cape, 1998.

Friedberg, Anne. *The Virtual Window: From Alberti to Microsoft*. Cambridge, MA: MIT Press, 2006.

Fuller, Matthew, and Andrew Goffey. *Evil Media*. Cambridge, MA: MIT Press, 2012.

Galison, Peter. *Image and Logic: A Material Culture of Microphysics*. Chicago: University of Chicago Press, 1997.

Galloway, Alexander. *The Interface Effect*. Cambridge: Polity, 2012.

Gentner, W., H. Maier-Leibnitz, and W. Bothe. *An Atlas of Typical Expansion Chamber Photographs*. London: Pergamon Press, 1954.

Gibbs, Samuel. "Microsoft's Racist Chatbot Returns with a Drug-Smoking Twitter Meltdown." *Guardian* (London), March 30, 2016.

Gillespie, Tarleton. "Algorithm." In *Digital Keywords: A Vocabulary of Information Society and Culture*, edited by Benjamin Peters, 18–30. Princeton, NJ: Princeton University Press, 2016.

Graham, Stephen. *Cities under Siege*. London: Verso, 2011.

Graham, Stephen. *Vertical: The City from Above and Below*. London: Verso, 2016.

Greenfield, Patrick. "The Cambridge Analytica Files: The Story So Far." *Guardian* (London), March 25, 2018.

Greenwald, Glenn. *No Place to Hide: Edward Snowden, the NSA, and the Surveillance State*. New York: Random House, 2014.

Gregory, Derek. "Drone Geographies." *Radical Philosophy* 183 (2014): 7–19.

Grothoff, Christian, and Jens Porup. "The NSA's SKYNET Program May Be Killing Thousands of Innocent People." *Ars Technica*, February 16, 2016. http://arstechnica.co.uk/security/2016/02/the-nsas-skynet-program-may-be-killing-thousands-of-innocent-people.

Grusin, Richard. *Premediation: Affect and Mediality after 9/11*. London: Palgrave Macmillan, 2010.

Guizzo, Erico. "Google Wants Robots to Acquire New Skills by Learning from Each Other." *IEEE Spectrum*, October 5, 2016.

Halpern, Orit. *Beautiful Data: A History of Vision and Reason since 1945*. Durham, NC: Duke University Press, 2014.

Halpern, Paul. *The Quantum Labyrinth: How Richard Feynman and John Wheeler Revolutionized Time and Reality*. New York: Basic Books, 2017.

Hansen, Mark. *New Philosophy for New Media*. Cambridge, MA: MIT Press, 2006.

Haraway, Donna. *Modest_Witness@Second_Millennium. FemaleMan©_Meets_ Onco-mouse.* New York: Routledge, 1997.

Haraway, Donna. "Situated Knowledges: The Science Question in Feminism and the Privilege of Partial Perspective." *Feminist Studies* 14, no. 3 (1988): 575–99.

Haraway, Donna. *Staying with the Trouble: Making Kin in the Chthulucene.* Durham, NC: Duke University Press, 2016.

Harding, Luke. *The Snowden Files.* London: Faber and Faber, 2014.

Hayes, Brian. "Cloud Computing." *Communications of the ACM* 51, no. 7 (2008): 9–11.

Hayles, N. Katherine. "How We Became Posthuman: Ten Years on, an Interview with N. Katherine Hayles." *Paragraph* 33, no. 3 (2010): 318–30.

Hayles, N. Katherine. *How We Became Posthuman: Virtual Bodies in Cybernetics, Literature, and Informatics.* Chicago: University of Chicago Press, 1999.

Hayles, N. Katherine. *How We Think: Digital Media and Contemporary Technogenesis.* Chicago: University of Chicago Press, 2012.

Hayles, N. Katherine. *My Mother Was a Computer: Digital Subjects and Literary Texts.* Chicago: University of Chicago Press, 2005.

Hayles, N. Katherine. *Unthought: The Power of the Cognitive Nonconscious.* Chicago: University of Chicago Press, 2017.

Heidegger, Martin. "Bauen, Wohnen, Denken." In *Vorträge und Aufsätze*, 145–62. Pfullingen: Neske, 1954.

Hendrycks, Dan, Kevin Zhao, Steven Basent, Jacob Steinhardt, and David Song. "Natural Adversarial Examples." arXiv:1907-07174V1 (2018): 2–16.

Hinton, Geoffrey. "Deep Neural Networks for Acoustic Modeling in Speech Recognition." *IEEE Signals Processing* 29, no. 6 (2012): 82–97.

Hinton, Geoffrey, Li Deng, Dong Yu, George Dahl, Abdel-Rahman Mohames, Navdeep Jaitly, Andrew Senior, Vincent Vanhoucke, Patrick Nguyen, Tara Sainath, and Brian Kingsbury. "Deep Neural Networks for Acoustic Modeling in Speech Recognition." *IEEE Signal Processing Magazine* 82 (2012): 83.

Hinton, Geoffrey, Simon Osindero, and Yee Whye Teh. "A Fast Learning Algorithm for Deep Belief Networks." *Neural Computation* 18, no. 7 (2006): 1527–54.

Hinton, Geoffrey, and Ruslan Salakhutdinov. "Reducing the Dimensionality of Data with Neural Networks." *Science* (2006): 504–7.

Hirschburg, J., and C. D. Manning. "Advances in Natural Language Processing." *Science* 349 (2015): 261–66.

Holzmann, G. J. "Points of Truth." *IEEE Software*, July/August 2015.

House of Commons Science and Technology Select Committee. *Algorithms in Decision Making.* London: HMSO, 2018.

House of Representatives Committee on Science and Technology. *Investigation of the Challenger Accident.* H. R. Rep. No. 99-1016. 99th Cong., 2d sess. Washington, DC: GPO. https://www.gpo.gov/fdsys/pkg/GPO-CRPT-99hrpt1016/pdf/GPO-CRPT-99hrpt1016.pdf.

Hu, Tung-Hui. *A Prehistory of the Cloud.* Cambridge, MA: MIT Press, 2015.

Intelligence and Security Committee of Parliament. *Privacy and Security: A Modern and Transparent Legal Framework.* London: HMSO, 2015.

204

Bibliography

Intelligence and Security Committee of Parliament. "Transcript of Evidence Given by the Rt. Hon. Philip Hammond MP (Foreign Secretary)." Privacy and Security Inquiry, October 23, 2014. http://isc.independent.gov.uk/public-evidence.

Jaeger, Paul. "Where Is the Cloud? Geography, Economics, Environment and Jurisdiction." *First Monday* 14, no. 5 (2009). https://firstmonday.org/ojs/index.php/fm/article/view/2456/2171.

Jay, Martin. "Cultural Relativism and the Visual Turn." *Journal of Visual Culture* 1, no. 3 (2002): 267–78.

Jefferson, Geoffrey. "The Mind of Mechanical Man." *British Medical Journal* 1, no. 4616 (1949): 1105–10.

Jochum, Elizabeth, and Ken Goldberg. "Cultivating the Uncanny: The Telegarden and Other Oddities." In *Robots and Art*, edited by Damith Herath and Christian Kroos, 149–75. New York: Springer, 2016.

Josephson, J., and S. Josephson. *Abductive Inference: Computation, Philosophy, Technology.* Cambridge: Cambridge University Press, 1996.

Juola, Patrick. "How a Computer Program Helped Show J. K. Rowling Wrote a Cuckoo's Calling." *Scientific American*, August 20, 2013.

Kant, Immanuel. *Critique of Pure Reason.* Cambridge: Cambridge University Press, 1998.

Kant, Immanuel. *Groundwork of the Metaphysics of Morals.* Translated by H. J. Patton. New York: Harper, 1964.

Keenan, Thomas. *Fables of Responsibility: Aberrations and Predicaments in Ethics and Politics.* Stanford, CA: Stanford University Press, 1997.

Kinsley, Sam. "The Matter of Virtual Geographies." *Progress in Human Geography* 38, no. 3 (2014): 364–84.

Kirby, Vicki. *Quantum Anthropologies: Life at Large.* Durham, NC: Duke University Press, 2011.

Kirchner, Lauren. "Federal Judge Unseals New York Crime Lab's Software for Analyzing DNA Evidence." *ProPublica*, October 20, 2017.

Kirchner, Lauren. "New York City Moves to Create Accountability for Algorithms." *Ars Technica*, December 19, 2017.

Kirchner, Lauren. "Traces of Crime: How New York's DNA Techniques Became Tainted." *New York Times*, September 4, 2017.

Kitchin, Rob. *The Data Revolution: Big Data, Open Data, Data Infrastructures and Their Consequences.* London: Sage, 2014.

Kittler, Friedrich. *Discourse Networks 1800/1900.* Stanford, CA: Stanford University Press, 1990.

Kittler, Friedrich. *Literature, Media, Information Systems.* Amsterdam: G + B Arts International, 1997.

Koehn, Philipp. "Europarl: A Parallel Corpus for Statistical Machine Translation." Machine Translation Summit, Phuket, Thailand, September 12–16, 2005.

Konkel, F. "The CIA's Deal with Amazon." *Atlantic*, July 17, 2014.

Kourou, Konstantina, Themis P. Exarchos, Konstantinos P. Exarchos, Michalis V. Karamouzis, and Dimitrios I. Fotiadis. "Machine Learning Applications in Cancer Prognosis and Prediction." *Computational and Structural Biotechnology Journal* 13 (2015): 8–17.

Krizhevsky, Alex, Ilya Sutskever, and Geoffrey Hinton. "ImageNet Classification with Deep Convolutional Neural Networks." *Advances in Neural Information Processing Systems* 2 (2012): 1097–1105.

Kurzweil, Ray. *The Singularity Is Near: When Humans Transcend Biology.* London: Viking, 2005.

Lange, Ann-Christina, Marc Lenglet, and Robert Seyfert. "Cultures of High-Frequency Trading: Mapping the Landscape of Algorithmic Developments in Contemporary Financial Markets." *Economy and Society* 45 (2016): 149–65.

Latour, Bruno, and Peter Weibel. *Making Things Public: Atmospheres of Democracy.* Cambridge, MA: MIT Press, 2005.

Laughland, Oliver. "Baltimore Unrest: 49 Children Were Arrested and Detained during Protests." *Guardian* (London), May 7, 2015.

LeCun, Yann, Yoshua Bengio, and Geoffrey Hinton. "Deep Learning." *Nature* 521, no. 28 (2015): 436–44.

Lemov, Rebecca. *Database of Dreams: The Lost Quest to Catalog Humanity.* New Haven, CT: Yale University Press, 2015.

Leopold, G. "CIA Embraces Cloudera Data Hub." *EnterpriseTech*, February 26, 2015.

Lewis, Paul. "Fiction Is Outperforming Reality: How YouTube's Algorithm Distorts the Truth." *Guardian* (London), February 3, 2018.

Lin, Henry, Izhak Shafran, Todd Murphy, Allison Okamura, David Yuh, and Gregory Hager. "Automatic Detection and Segmentation of Robot-Assisted Surgical Motions." *Medical Image Computing* 8, no. 1 (2005): 802–10.

Liptak, Adam. "Sent to Prison by a Software Program's Secret Algorithms." *New York Times*, May 1, 2017. https://www.nytimes.com/2017/05/01/us/politics/sent-to-prison-by-a-software-programs-secret-algorithms.html.

Liu, Lydia H. *The Freudian Robot: Digital Media and the Future of the Unconscious.* Chicago: University of Chicago Press, 2010.

"Machine Learning for Cybersecurity." Cloudera. Accessed April 2018. https://www.cloudera.com/solutions/gallery/digital-reasoning-synthesys.html.

MacKay, David J. C. *Information Theory, Inference, and Learning Algorithms.* Cambridge: Cambridge University Press, 2017.

Mackenzie, Adrian. *Machine Learners: Archaeology of a Data Practice.* Cambridge, MA: MIT Press, 2017.

MacKenzie, Donald. "How Algorithms Interact: Goffman's 'Interaction Order' in Automated Trading." *Theory, Culture and Society* 36, no. 2 (2019): 39–59.

MacKenzie, Donald. "How to Make Money in Microseconds." *London Review of Books* 33, no. 10 (2011): 16–18.

Mahler, J., J. Liang, S. Niyaz, M. Laskey, R. Doan, X. Liu, J. A. Ojea, and K. Goldberg. "Dex-Net 1.0: A Cloud-Based Network of 3D Objects for Robust Grasp Planning Using a Multi-Armed Bandit Model with Correlated Rewards." Paper presented at the International Conference on Robotics and Automation, Stockholm, May 16–21, 2016.

Manovich, Lev. *The Language of New Media*. Cambridge, MA: MIT Press, 2002.

Mantel, Hilary. "The Day Is for the Living: Hilary Mantel on Writing Historical Fiction." BBC Reith Lectures, June 13, 2017. *Medium*. https://medium.com /@bbcradiofour/hilary-mantel-bbc-reith-lectures-2017-aeff8935ab33.

Markoff, John. "Artificial Intelligence Is Far from Matching Humans, Panel Says." *New York Times*, May 26, 2016. https://www.nytimes.com/2016/05/26/technology /artificial-intelligence-is-far-from-matching-humans-panel-says.html.

Markov, A. A. *Theory of Algorithms*. Translated by Jacques Schorr-Kon. Moscow: Academy of Sciences of the USSR, 1954.

Mattern, Shannon. *Code and Clay, Data and Dirt: Five Thousand Years of Urban Media*. Minneapolis: University of Minnesota Press, 2017.

Mayer-Schönberger, Viktor, and Kenneth Cukier. *Big Data: A Revolution That Will Change How We Live, Work, and Think*. New York: Houghton Mifflin Harcourt, 2013.

McCarthy, John. "Architects of the Information Society: Thirty-Five Years of the Laboratory for Computer Science at MIT." Lecture at the MIT Centennial, July 1961.

McCormack, Derek. "Remotely Sensing Affective Afterlives: The Spectral Geographies of Material Remains." *Annals of the Association of American Geographers* 100, no. 3 (2010): 640–54.

McGeoch, Catherine C. *A Guide to Experimental Algorithmics*. Cambridge: Cambridge University Press, 2012.

Mehotra, Kartikay. "Maker of Surgical Robot Hid More than 700 Injury Claims, Insurer Says." *Bloomberg*, May 26, 2016.

"Memorandum in Support of Application by ProPublica for Leave to Intervene, Lift the Protective Order and Unseal Judicial Records." Filed September 25, 2017. Yale Law School website, https://law.yale.edu/system/files/area/clinic/document/139 ._memorandum_of_law.pdf.

Metcalf, Jacob, Emily F. Keller, and danah boyd. "Perspectives on Big Data, Ethics, and Society." Report for the Council for Big Data, Ethics, and Society, May 23, 2016. https://bdes.datasociety.net/wp-content/uploads/2016/05/Perspectives-on -Big-Data.pdf.

Misra, C., P. K. Swain, and J. K. Mantri. "Text Extraction and Recognition from Image Using Neural Network." *International Journal of Computer Applications* 40, no. 2 (2012): 13–19.

Mitchell, David. *Cloud Atlas*. London: Hodder and Stoughton, 2004.

Mitchell, W. J. T. "There Are No Visual Media." *Journal of Visual Culture* 4, no. 2 (2005): 257–66.

Mol, Anne Marie. *The Body Multiple: Ontology in Medical Practice*. Durham, NC: Duke University Press, 2002.

"Monitoring Protests from Unstructured Text." 29:54, YouTube, 2012. https://www .youtube.com/watch?v=6Sbny9iNjeA.

National Research Council of the National Academies. *Bulk Collection of Signals Intelligence: Technical Options*. Washington, DC: National Academies Press, 2015.

Nature Editors. "More Accountability for Big Data Algorithms." *Nature* 537, no. 7621

(2016): 449. https://www.nature.com/news/more-accountability-for-big-data
-algorithms-1.20653.

Newman, M. H. A., A. Turing, G. Jefferson, and R. B. Braithwaite. "Can Automatic
Calculating Machines Be Said to Think?" Transcript of discussion, BBC, recorded
1951, aired January 14, 23, 1952. http://www.turingarchive.org/browse.php/b/6.

Nissenbaum, Helen. *Privacy in Context: Technology, Policy, and the Integrity of Social Life.*
Stanford, CA: Stanford University Press, 2009.

Noble, Safiya. *Algorithms of Oppression: How Search Engines Reinforce Racism.* New York:
NYU Press, 2018.

NSF Award Abstract 1734521. Last amended October 27, 2017. https://www.nsf.gov
/awardsearch/showAward?AWD_ID=1734521&HistoricalAwards=false.

Nyholm, Sven, and Jilles Snids. "The Ethics of Accident-Algorithms for Self-Driving
Cars." *Ethical Theory and Moral Practice* 19, no. 5 (2016): 1275–89.

O'Brien, John. *The Definitive Guide to the Data Lake.* Boulder, CO: Radiant Advisors, 2015.

Office of the Director of National Intelligence. "ICITE Fact Sheet." Washington, DC:
ODNI, 2012.

Office of the Director of National Intelligence. *Intelligence Community Information
Technology Enterprise Strategy 2016-2020.* Washington, DC: ODNI, 2015.

Ohri, A. *R for Cloud Computing: An Approach for Data Scientists.* London: Springer, 2014.

O'Neil, Cathy. *Weapons of Math Destruction: How Big Data Increases Inequality and
Threatens Democracy.* New York: Crown, 2016.

Ormandy, Roman. "Google and Elon Musk to Decide What Is Good for Humanity."
Wired, January 26, 2015.

Paglen, Trevor. *Invisible: Covert Operations and Classified Landscapes.* New York: Aperture, 2010.

Paglen, Trevor. "Trevor Paglen Offers a Glimpse of America's Vast Surveillance Infrastructure." American Suburb X, December 30, 2014. http://www.american
suburbx.com/2014/12/trevor-paglen-offers-a-glimpse-of-americas-vast-surveillance
-infrastructure.html.

Parikka, Jussi. *A Geology of Media.* Minneapolis: University of Minnesota Press, 2015.

Parikka, Jussi. *Insect Media: An Archaeology of Animals and Technology.* Minneapolis: University of Minnesota Press, 2010.

Parisi, Luciana. *Contagious Architecture: Computation, Aesthetics, and Space.* Cambridge,
MA: MIT Press, 2013.

Parkin, Simon. "Teaching Robots Right from Wrong." *Economist 1843 Magazine,* June/
July 2017. https://www.1843magazine.com/features/teaching-robots-right
-from-wrong.

Parks, Lisa, and Caren Kaplan. *Life in the Age of Drone Warfare.* Durham, NC: Duke
University Press, 2017.

Pasquale, Frank. *The Black Box Society: The Secret Algorithms That Control Money and Information.* Cambridge, MA: Harvard University Press, 2015.

Peters, John Durham. *The Marvelous Clouds: Toward a Philosophy of Elemental Media.*
Chicago: University of Chicago Press, 2015.

Phys.org. "UMass Lowell Professor Steers Ethical Debate on Self-Driving Cars." *UMass Lowell News*, October 5, 2017. https://www.uml.edu/News/news-articles/2017/phys-self-driving-ethics.aspx.

Pickett, Marc, Chris Tar, and Brian Strope. "On the Personalities of Dead Authors." *Google AI Blog*, February 24, 2016. https://ai.googleblog.com/2016/02/on-personalities-of-dead-authors.html.

Pishko, Jessica. "The Impenetrable Program Transforming How Courts Treat DNA Evidence." *Wired*, November 29, 2017.

Popescu, M. C., and N. Mastorakis. "Simulation of da Vinci Surgical Robot Using Mobotsim Program." *International Journal of Biology and Biomedical Engineering* 4, no. 2 (2008): 137–46.

Prentice, Rachel. *Bodies in Formation: An Ethnography of Anatomy and Surgery Education*. Durham, NC: Duke University Press, 2013.

Puar, Jasbir. *Terrorist Assemblages: Homonationalism in Queer Times*. Durham, NC: Duke University Press, 2007.

Quinlan, J. R. "Induction of Decision Trees." *Machine Learning* 1, no. 1 (1986): 81–106.

Reiley, Carol, Erion Plaku, and Gregory Hager. "Motion Generation of Robotic Surgical Tasks: Learning from Expert Demonstrations." Paper presented at the Annual International Conference of the IEEE Engineering in Medicine and Biology, Buenos Aires, Argentina, August 31–September 4, 2010.

Rochester, G. D., and J. G. Wilson. *Cloud Chamber Photographs of the Cosmic Radiation*. London: Pergamon Press, 1952.

Rose, Gillian. "Rethinking the Geographies of Cultural 'Objects' through Digital Technologies: Interface, Network, Friction." *Progress in Human Geography* 40, no. 3 (2015): 334–51.

Rotman, Brian. *Becoming Beside Ourselves: The Alphabet, Ghosts, and Distributed Human Being*. Durham, NC: Duke University Press, 2008.

Said, Edward. *On Late Style: Music and Literature against the Grain*. London: Bloomsbury, 2007.

Schneier, Bruce. "NSA Robots Are Collecting Your Data, Too, and They're Getting Away with It." *Guardian* (London), February 27, 2014.

Scorer, Richard, and Harry Wexler. *Cloud Studies in Colour*. Oxford: Pergamon Press, 1967.

Seaver, Nick. "Algorithms as Culture: Some Tactics for the Ethnography of Algorithmic Systems." *Big Data and Society* 3, no. 2 (2017): 2.

Shane, Scott, and Daisuke Wakabayashi. "The Business of War: Google Employees Protest Work for the Pentagon." *New York Times*, April 4, 2018.

Sharkey, Noel. "Staying in the Loop: Human Supervisory Control of Weapons." In *Autonomous Weapons Systems: Law, Ethics, Policy*, edited by Nehal Bhuta, Susanne Beck, Robin Geiss, Hin-Yan Liu, and Claus Kress, 23–38. Cambridge: Cambridge University Press, 2016.

Shaw, Ian, and Majed Akhter. "The Dronification of State Violence." *Critical Asian Studies* 46, no. 2 (2014): 211–34.

Simonyan, Karen, and Andrew Zisserman. "Very Deep Convolutional Networks for

Large-Scale Image Recognition." *arXiv:1409.1556* (September 2014). https://arxiv
.org/abs/1409.1556v1.

Stengers, Isabelle. *Cosmopolitics I.* Minneapolis: University of Minnesota Press, 2010.

Stengers, Isabelle. *Thinking with Whitehead: A Free and Wild Creation of Concepts.* Cambridge, MA: Harvard University Press, 2011.

Strawser, B. J. *Killing by Remote Control: The Ethics of an Unmanned Military.* Oxford: Oxford University Press, 2013.

Stubb, Alexander. "Why Democracies Should Care Who Codes the Algorithms." *Financial Times*, June 2, 2017.

Suchman, Lucy, and Jutta Weber. "Human-Machine Autonomies." In *Autonomous Weapons Systems: Law, Ethics, Policy*, edited by N. Bhuta, Susanne Beck, Robin Geiss, Hin-Yan Liu, and Claus Kress, 75–102. Cambridge: Cambridge University Press, 2016.

Taylor, Duncan A., Jo-Anne Bright, and John Buckleton. "A Source of Error: Computer Code, Criminal Defendants, and the Constitution." *Frontiers in Genetics* 8 (2017): 33.

Timmer, John. "Software Faults Raise Questions about the Validity of Brain Studies." *Ars Technica*, July 1, 2016.

Totaro, Paulo, and Domenico Ninno. "The Concept of Algorithm as an Interpretative Key of Modern Rationality." *Theory, Culture and Society* 31, no. 4 (2014): 32.

Truvé, Staffan. "Building Threat Analyst Centaurs Using Artificial Intelligence." Recorded Future, January 14, 2016. https://www.recordedfuture.com/artificial -threat-intelligence/.

Tucker, Patrick. "A New AI Learns through Observation Alone." *Defense One*, September 6, 2016.

Turing, Alan. "Letters on Logic to Max Newman." In *The Essential Turing*, edited by B. Jack Copeland, 205–16. Oxford: Oxford University Press, 2004.

Turing, Alan. "Systems of Logic Based on Ordinals." *Proceedings of the London Mathematical Society* 2, no. 45 (1939): 161–228.

Turkle, Sherry. *The Second Self: Computers and the Human Spirit.* Cambridge, MA: MIT Press, 2005.

van den Berg, J., S. Miller, D. Duckworth, H. Hu, A. Wan, X. Fu, K. Goldberg, and P. Abbeel. "Superhuman Performance of Surgical Tasks by Robots Using Iterative Learning." Paper presented at the IEEE International Conference on Robotics and Automation, Anchorage, Alaska, May 3–7, 2010.

Vaughan, Diane. *The Challenger Launch Decision: Risky Technology, Culture and Deviance at NASA.* Cambridge, MA: MIT Press, 1996.

Vinyals, Oriol, Alexander Toshev, Samy Bengio, and Dumitru Erhan. "Show and Tell: A Neural Image Caption Generator." *IEEE Transactions on Pattern Analysis and Machine Intelligence* 39, no. 4 (2017): 2048–57.

Walker, James. "Researchers Shut Down AI That Invented Its Own Language." *Digital Journal*, July 21, 2017. http://www.digitaljournal.com/tech-and-science /technology/a-step-closer-to-skynet-ai-invents-a-language-humans-can-t-read /article/498142.

Weber, Jutta. "Keep Adding: On Kill Lists, Drone Warfare, and the Politics of Databases." *Environment and Planning D: Society and Space* 34, no. 1 (2016).

Weber, Samuel. *Targets of Opportunity: On the Militarization of Thinking.* New York: Fordham University Press, 2005.

"Web Intelligence." Recorded Future. Accessed January 2018. https://www.recorded future.com/web-intelligence/.

Webster, Katharine. "Philosophy Professor Wins NSF Grant on Ethics of Self Driving Cars." *UMass Lowell News*, August 8, 2017. https://www.uml.edu/news/stories/2017/selfdrivingcars.aspx.

Weizman, Eyal. "The Politics of Verticality: The West Bank as an Architectural Construction." In *Territories: Islands, Camps, and Other States of Utopia*, edited by A. Franke, 106–24. Cologne: Walther Koenig, 2004.

Wiener, Norbert. "A Scientist Rebels." *Atlantic Monthly*, January 1947.

Wilcox, Lauren. "Embodying Algorithmic War: Gender, Race, and the Posthuman in Drone Warfare." *Security Dialogue* 48, no. 1 (2017): 11–28.

Wilson, C. T. R. "On the Cloud Method of Making Visible Ions and the Tracks of Ionizing Particles." In *Nobel Lectures in Physics, 1922–1941*, 167–217. Amsterdam: Elsevier, 1927.

Wilson, J. G. *The Principles of Cloud Chamber Technique.* Cambridge: Cambridge University Press, 1951.

Winslett, Marianne. "Rakesh Agrawal Speaks Out." Interview with Rakesh Agrawal. *SIGMOD Record* 32, no. 3 (2003). https://sigmod.org/publications/interviews/pdf/D15.rakesh-final-final.pdf.

Wittgenstein, Ludwig. *On Certainty.* Edited by G. E. M. Anscombe and G. H. von Wright. Oxford: Blackwell, 1975.

Wittgenstein, Ludwig. *Philosophical Investigations.* Translated by G. E. M. Anscombe, P. M. S. Hacker, and Joachim Schulter. Oxford: Wiley Blackwell, 2008.

Woodman, Spencer. "Palantir Provides the Engine for Donald Trump's Deportation Regime." *Intercept*, March 2, 2017.

Xu, K., J. Lei Ba, R. Kiros, K. Cho, A. Courville, R. Salakhutdinov, R. Zemel, and Y. Bengio. "Show, Attend and Tell: Neural Image Caption Generation and Visual Attention." Paper presented at the 32nd International Conference on Machine Learning, Lille, France, July 6–11, 2015.

Yang, A. Y., X. Xhou, A. G. Balasubramanian, S. S. Sastry, and Y. Ma. "Fast ℓ_1-Minimization Algorithms for Robust Facial Recognition." *Transactions on Image Processing* 22, no. 8 (2013): 3234–46.

Zhang, Q., L. Cheng, and R. Boutaba. "Cloud Computing: State-of-the-Art and Challenges." *Journal of Internet Services and Applications* 1 (2010): 7–18.

aberration, 108–29

abduction, 47–48, 75, 161–62

abnormality, 6, 43, 47, 68, 78–79, 92, 97, 115, 139. *See also* norms

accidents, 75–76, 115–16, 119

accountability: algorithms and, 5, 11, 18–20, 38, 85, 87, 166; episteme of, 8–9; in robot surgery, 66; of science, 134

accounts: giving of, 18–20, 62, 80, 135, 140–48, 152, 164–68

action, 16, 19, 41, 46, 109–10; theory of as a "spring," 59–60, 68, 76–77

AlexNet, 72–74, 77, 79, 136

algorithms: accounts of themselves, 18–20, 54, 72, 102, 135, 148, 171; as aperture instruments, 15–18, 40, 156; as arrangements of propositions, 6–7, 9, 11–14, 40, 57, 87, 96, 99, 104, 136, 145, 158, 162; as composites, 10, 20, 42, 57, 64, 66, 88, 135; deep learning, 3, 89, 136, 150; designers of, 18, 67–68, 139, 147; and experimentation, 12, 40–41, 43, 48, 53, 67–68, 92, 118–19, 139, 161; and layers, 13, 35; and modification, 11, 48, 74, 87, 96–97, 106, 137; and multiplicity, 13, 17, 64, 80, 87, 121, 152, 162; and steps, 11, 13, 40, 69, 89

Amazon Web Services (AWS), 33, 35, 49

animals, 118, 137, 157

anomaly detection, 40, 42–43, 68, 79. *See also* patterns

apertures, 15–18, 42–44, 46, 48, 71, 94, 104–5, 160–64

application programming interface (API), 59

applications, 35, 49

architecture, 14, 33, 35, 64, 111, 136

archiving, 49–51

art, 5

artificial intelligence (AI), 109, 146

associative life, 7

atomic weapons, 112, 133

attention, 15, 42, 154–55, 157, 162

attributes: and attribution, 41, 85–107, 118; and clustering of data, 1, 4, 33, 48–49, 70, 122, 126, 160; racialized, 11; and selves and others, 8, 20, 43, 58, 79, 138, 148; and the unattributable, 25, 154–72

author: algorithm designer as, 18; and authorship, 85–107, 135

automation, 15, 18, 33, 85

autonomous machines, 57, 65, 90, 109

autonomous vehicles, 76, 119–22

autonomous weapons, 11, 66, 175n33, 182n30

back propagation, 66, 106

Barad, Karen: on algorithms, 141; on intelligibility, 42; on mattering, 10, 46, 144, 152; on touch, 63

Bergson, Henri, 16, 42, 152, 163

Berlant, Lauren, 147

bias: algorithmic, 5, 8, 18–19, 38, 87, 92, 95–96, 146; and gender, 11; and machine learning, 74–75, 79, 81, 106, 125

bifurcation, 88, 98–99, 113

big data, 16, 18, 34–35, 77. *See also* data

biometrics, 3, 4, 14, 65, 70, 135, 147. *See also* facial recognition

black box, 5, 40, 73, 110

borders, 33, 46–47, 54, 67–68, 112, 155; and immigration, 12, 55, 139, 161

Bratton, Benjamin, 35, 40

Breiman, Leo, 125

Browne, Simone, 4

Butler, Judith, 8, 19, 38, 135, 145, 151

calculation: algorithmic forms of, 11, 21,
33, 42, 46, 75, 95, 112; archiving of data
for, 50, 54; and decision-making, 120,
128, 140–41, 143, 150; and incalculability, 162; incorporation of doubt in, 134,
142, 151; probabilistic methods of, 125
Cambridge Analytica, 4–5, 118, 160
cancer, 123
categorization, 14
Central Intelligence Agency (CIA), 31,
33, 50
Challenger (space shuttle), 140–53
cities, 1–4, 20, 85, 155
classification, 33, 40, 49, 92, 115–19; and
classifiers, 9, 11, 14, 68, 70, 125, 139;
and clouds, 29
cloud atlas, 44, 48–49
Cloud Atlas (Mitchell), 55
cloud chamber, 29–31, 44, 48, 50, 52–53
cloud computing, 7, 30–31, 33, 35, 38,
40–42, 46–47
cloud data: algorithmic analysis of, 7,
42, 49, 62, 79; architectures of, 21, 70,
72, 76–77; medical images as, 59; storage of, 37
Cloudera, 33
cloud ethics: and doubt, 135, 137–38; and
encoded ethics, 121, 123; and partial
account, 141; and political claims,
80–81, 146, 148; theory of, 7–8, 17, 19,
88, 117, 154–72. *See also* ethics
cloud forms, 40
cloud imaginations, 36
cloud robotics, 76
clustering: in anomaly detection, 40,
42–44, 92, 122, 136; of data for protest
detection, 1–4, 169; in elections, 161;
and false positives, 128; mapping of
features via, 58, 70, 155; in natural language processing, 89–90; and predictions, 138, 173n9
code, 1, 7, 9, 40, 59; source, 18, 85–88,
91–92, 95–96, 99, 104, 186n34. *See also*
writing

Cold War, 42, 113–14, 140
communication, 37
computation, 7, 34–35, 44, 137, 139, 165
computer science, 11, 34–35, 71–72,
88, 93, 104, 106, 113, 137, 135. *See also*
science
condensation, 16, 29–30, 43, 44–46, 49,
77, 99, 139, 156, 162
contingency, 21, 68, 98, 146
control, 58, 65, 79, 113
correlation, 31, 40, 44, 47, 50, 107, 138,
161
counterterrorism. *See* terrorism
Crary, Jonathan, 15, 39, 162
credit scoring, 18
criminal justice, 11, 85, 95, 104, 137
Critchley, Simon, 100, 106, 171
crowd: algorithms to assess risk in, 4, 17,
155, 159–60; crowded court, 25, 69
cybernetics, 111–15, 140

Daston, Lorraine, 125
data: and algorithms, 7, 18, 49, 50, 86,
88, 92, 96, 134–35, 138, 141–53; and features, 12, 16, 20, 43, 155, 160; sharing
of, 33, 42, 50; social media and, 1–2,
37, 42, 47, 89; storage of, 36, 38; unlabeled, 90. *See also* big data
data analysis, 31, 33
data centers, 33–35, 38–39
data environment, 58, 75
data inputs, 11, 48, 58, 90, 105, 125, 127, 160
data lake, 47
data mining, 36, 179n63
data residue, 46, 59, 64, 164
da Vinci robot, 59, 63–64
decision procedures, 42, 57, 69, 136, 139
decisions: and automation, 11, 57, 67, 85,
102; doubt and, 134, 137, 150; human,
22, 62, 65–66, 140–48; and machine
learning, 12, 71, 80, 87, 92, 106; madness of, 112–14, 119–23; and undecidability, 19, 20, 88, 98–99, 128, 151–53,
162–65

decision trees, 10, 70, 89 112–113, 125, 165.
See also random forests
deconstruction, 88, 99, 104
deep learning. See machine learning;
neural networks
Deleuze, Gilles: on duration, 17, 163; on
ethics, 165; on fabulation, 98, 102, 158,
161–62
democracy, 4, 38, 118, 160
derivatives, 50, 106, 124–25
Derrida, Jacques: on archives, 51; on the
author, 99; on calculation, 128; on decision, 121, 149, 152; on ethics, 166, 169;
on madness, 110, 116; on nonclosure,
104, 154, 156; on text, 104
DexNet, 77
the digital, 35
digital subjects, 94
discrimination, 8, 38, 85
DNA, 95–97, 105
doubt, 9, 10, 13, 133–53
drones, 33, 47, 78, 80, 124–26, 176n55;
image recognition systems and, 16,
38, 111, 146, 156, 160; targeting of, 55,
127, 128

economy, 35, 43, 46, 49, 89, 108, 123
embodiment, 57, 59, 62, 64, 137–38, 143
entanglement, 16, 57, 59
error and errancy, 47–48, 108–12, 115,
116, 120–23, 136, 146–47; experimentation and, 74–75, 96, 118, 125, 159,
169; in robot surgery, 58; sources of,
20, 86
ethicopolitics: of algorithms, 7–8, 10–11,
47, 53, 58, 71, 87, 88, 110, 146, 156, 165,
171–72; bias and, 75, 79; materiality
and, 36; theories of, 19–20, 33, 65–66,
69, 121; writing and, 22, 92, 105. See
also politics
ethics: codes of, 5–7, 15, 19, 86, 95, 110,
119, 120–21, 128, 146; of deconstruction, 104, 106; fabulation and, 98–99;
machine learning and, 55–56, 66, 80,

88, 90, 105, 111, 118, 135, 147–53, 154–72;
and opacity, 5, 14. See also cloud ethics
European Union, 36
experimentation, 12, 30, 40–41, 43–44,
48, 53, 67, 90, 161
expertise, 59–60, 96
explainability, 18
extraction. See feature extraction

fabulation, 98–99, 102–5, 158, 160
Facebook, 37, 50, 89, 138, 160
facial recognition, 14, 48, 68–70, 135–36.
See also recognition
false positive, 122, 145
fascism, 112
feature extraction, 14, 16, 58, 160; processes of, 42–43, 59, 63, 89–91, 139, 154
feature space, 58, 74, 79, 155, 160, 169,
181n8, 182n24
feature vectors, 59, 72, 155, 160, 167, 169,
171, 181n10
feedback loop, 66, 97, 106
feminism, 88, 19, 141
Feynman, Richard, 133–53
finance. See economy
forensics, 95
forking pathways, 88, 98–99, 118, 139,
141, 153, 161–62, 165
Foucault, Michel: on authorship, 18,
86–87; on ethics, 7, 10, 80–81, 171; on
madness, 115–18; on parrhesia, 145–47;
on probable events, 44; on truth-
telling, 5, 122
Fowles, John, 85, 88, 98–99, 101
fraud, 46, 53, 68, 91, 124
the future: algorithms and, 7, 17, 46,
49–51, 122; anticipation of, 72, 73;
inferring, 88–91; optimization of,
62, 80, 147; past data and, 20, 43,
46, 138; political claims on, 4, 54,
152, 153, 161

Galison, Peter, 30, 38, 40–41, 44, 49
gender, 85

geography: of cloud computing, 34–36, 38, 40; of data, 33
geopolitics, 33, 36–37, 114, 137
gestures, 59, 60, 62, 118
Github, 97
good: appearance of, 78–79, enough, 67, 69, 75, 161; and evil, 58, 109–10, 112, 134, 146, 158
Google, 17, 35, 77, 89–90, 92, 146, 154
government, 1, 33, 36, 42–43, 49, 113, 140, 149
grammar, 13
ground truth, 4, 133–53, 169. *See also* truth

Halpern, Orit, 15, 50, 54, 114
Haraway, Donna: on partiality, 20, 135, 137, 143, 166; on staying with the trouble, 19, 40, 66, 178n35, 182n31; on technoscience, 133–34
Hayles, N. Katherine, 137; on cognition, 19, 43; on computation, 4, 173n8; on data, 49; on the posthuman, 65–66
health, 123
Hinton, Geoffrey, 74
Homeland Security, US Department of (DHS), 2, 31, 43, 50
human, humans, 57, 60, 62, 64, 79, 88, 94; agency, 65, 112, 135; and algorithm relations, 9, 19–20, 42–43, 57, 63–64, 92, 139; labeling of images, 17, 72, 136, 148; in the loop, 11, 65–67, 109, 127, 152; subjects, 5, 118; visibility, 14, 30, 38. *See also* posthuman, posthumanism

ICITE, 31, 42, 49, 54
ImageNet, 17, 73, 173n7
image recognition, 14, 17, 70–71, 73, 121, 137. *See also* recognition
images, 30, 44, 48, 91, 155; drone, 16, 38; medical, 62–63, 122–23, 160
incalculability, 92, 110, 120, 141, 143, 155, 162

incompleteness, 21, 43, 80, 96
the incomputible, 64, 167
indeterminacy, 14, 87, 96, 156
inference, 43–44, 74, 89, 155; causality and, 47, 107, 122; of intent, 7, 91, 127; of motion, 46
infrastructure, 30, 33, 35, 39
insurance, 138
intelligence: geopolitical, 15, 31, 33, 54; surveillance and, 37–39, 42, 47, 126, 148
Intelligence Community Information Technology Enterprise (ICITE), 31, 42, 49, 54
internet, 30, 37, 50
intuition, 13, 57, 64, 67, 145, 150, 166
iteration, 11, 43, 92, 96–97, 139; and learning, 60, 62, 68, 106, 138

Keenan, Thomas, 19, 148–49, 152, 165
knowledge discovery, 47

language, 9, 43, 89. *See also* natural language processing
latency, 90
law, 37–38, 66, 95–96
layers, 13, 35, 62, 85; hidden, 71, 162
learning, 8–9, 56, 59, 90, 122. *See also* machine learning
likelihood. *See* probability
literature: authorship and, 87, 98–99, 102; as training data for machine learning, 89, 92–93
Los Alamos, New Mexico, 112, 133, 140

machine learning: anomaly detection and, 42–43, 53; deep, 89, 156–57; logics of, 14, 40, 87, 95–96, 118, 154, 157, 158, 160, 173n9; materiality and, 46–47; processes of, 8–9, 11, 20, 54, 56–81, 133–53, 173n7, 181n10; semi-supervised, 68, 124; and society, 5, 12, 109, 169, 170; unsupervised, 68–69, 111
Mackenzie, Adrian, 12

madness, 108–29
Mantel, Hilary, 102, 107
matching, 90
mathematics, 9, 13, 42, 56, 62, 94, 112, 140
Mechanical Turk, 72, 136, 148
media, 7, 15, 24, 59, 111, 178n38
Microsoft, 108–9
migration, 129
mistakes. *See* accidents; error and errancy
metadata. *See* data
models, 5, 59, 67–68, 89–90, 111, 142, 162
moral law, 6
moral panic, 16, 57, 110
moral philosophy, 119–20
moral responsibility. *See* responsibility
movement, 59, 62

National Security Agency (NSA), 31, 37, 39, 43, 50, 124
natural language processing (NLP), 50, 88–90, 92, 155–56, 167, 185n10, 185–86n12. *See also* language
nearest neighbor, 77
neural networks: convolutional, 14, 72, 77, 154–55, 160, 162, 166; deep, 10, 56–81, 89, 94, 105–6, 110, 114, 136, 137, 139, 161; drones and, 176n55; training of, 9, 16–17, 89, 111, 137, 192n13
New York, 85, 91, 93, 95
norms, 6, 13, 48, 136. *See also* abnormality

objectivity, 9, 20, 29, 106, 117
object recognition, 10, 16–17, 44, 69, 77–78, 146, 154. *See also* recognition
opacity: of algorithms, 5, 8, 19; Butler's conceptualization of, 135, 151; critical theory of, 23, 67, 98, 101, 105, 110, 117, 123, 140, 163, 164–68, 171–72; data and, 42, 136
optimization, 10, 12–13, 17, 43, 54, 60, 64, 147
output: algorithms and, 4, 17, 18, 47, 74–75, 80, 96, 97, 102, 104, 107, 111, 123, 128, 145, 155–56, 160, 162; decision and, 7; and optimization, 10, 12–13, 17, 43, 54, 60, 64, 147; and target, 9, 43, 67–68, 90, 105, 125, 139; and truth, 14, 114, 136–37. *See also* targets

Palantir, 4, 139, 147
parameters, 48, 62, 68, 80, 111, 121, 126, 161
Parisi, Luciana, 12, 47, 147
partiality, 8, 14, 123; Butler's conceptualization of, 19, 135; Haraway's conceptualization of, 20, 135; of images, 72; science and, 134; writing and, 9
particles, 30, 44, 48
pathways, 88, 98–99, 118, 139, 141, 153, 161–62, 165
patterns, 6, 17, 43–44, 46–49, 89, 126. *See also* anomaly detection
perception, 15–18, 29–30, 33, 40–41
performativity, 13, 104, 145
photography. *See* images
physics, 29, 40, 44, 48, 112, 134, 141
pixels, 70, 73–74, 160, 181n10
platform, 35, 49, 97, 109
play, 68, 71, 79, 80
point clouds, 75–76, 184n56
police, 2, 33, 94
politics, 4, 7, 8, 33, 36, 48, 55, 170; of algorithms, 7, 15, 42–43, 87, 139, 153, 160, 172; of appearance, 5, 15, 46, 72–73, 79, 155; and difficulty, 10, 152; and foreclosure, 20, 80, 93, 160–61; theory of, 19, 81, 104, 148–49, 150, 165, 167, 169, 171. *See also* ethicopolitics
possibility: of automation, 59; combinatorial, 14, 163; as computational layer, 137; of errors, 47; of future action, 51, 79, 97, 119, 147; perception and, 16, 46, 143
posthuman, posthumanism, 65, 88, 118, 135, 141, 142, 150–51. *See also* human, humans

potentiality, 7
pre-computation, 77–79, 120
prediction, 90, 106, 138, 160
privacy, 38
probability, 14, 48, 68, 70, 71; algorithms and, 74, 90, 92, 99, 105, 137, 162–63, 167; contingency of, 75, 80, 81, 139, 142–43, 147–48, 175–76n37; Daston's conceptualization of, 125; and the event, 44, 107; likelihood ratio and, 95–96; robotics and, 77
profiles, 69–70, 76, 95–96
PRISM, 37
propensities, 5, 17, 49, 91, 160, 194n15
protest, 1–4, 10, 17, 46, 55, 91, 169

race and racism, 1–4, 11, 69, 85, 97, 111, 128, 145, 175n37
random forests, 112, 124–27, 165, 190n51. See also decision trees
reason, 42, 44, 86, 108–29
recognition, 1, 3, 14, 17, 33; of authorship, 90; Butler's conceptualization of, 19, 145; of handwritten text, 69–70, 173n4; of patterns, 46, 150, 181n10; politics of, 4, 8, 41, 126, 127, 167, 171, 184n55; regimes of, 48–49, 54, 56–81, 121, 122, 183n43; semantic, 89, 185–86n12; speech, 106. See also facial recognition; image recognition; object recognition
reduction: algorithmic processes of, 16–17, 115, 128
resistance, 51, 145, 157, 169–70
responsibility: for algorithmic harms, 100, 112; and code, 86, 94, 95; ethics of, 8, 81, 118, 121, 123, 164–65; human touch and, 63–64, 144; locus of, 5, 18, 19, 66, 92, 99; of science, 114, 133–34
rights, 5, 10, 38, 148
risk, 4, 19, 40, 104, 126, 138, 140, 145–48
robots, 20, 58–59, 63–64, 77, 94
rules, 7, 14, 47, 53, 65, 69, 89, 141, 150

scene analysis, 16, 25, 43, 88, 126, 154, 157
science: apparatus of, 29–30, 34, 38, 44, 53; and knowledge, 16, 35, 87, 93, 98, 112, 133, 137–53; and observation, 29. See also computer science
SKYNET, 124–27
scores: algorithm-generated, 4, 68, 96, 126
security, 12, 40, 43, 50; agencies, 33, 37, 89; automation and, 140; military, 16, 112, 124–27; threats to, 54, 70, 74, 148; and uncertainty, 47, 134
sentiment analysis, 54, 147
sight, 15, 29, 31, 38, 40–41. See also vision
social media: chatbot and, 108–9; data, 1–2, 37, 42–43, 47, 50, 51, 71, 160, 170; images, 3, 54; sentiment analysis of, 147, 174n9; text, 70, 89, 91
Society for Worldwide Interbank Financial Telecommunications (SWIFT), 36
software, 35, 42, 97
source code, 18, 85–88, 91–92, 95–96, 99, 104, 186n34
sovereignty, 31, 36–38, 53–54, 128–29
spatiality, 11, 14, 33, 35, 37, 40, 71, 162–63
spoofing, 121
statistics, 12, 47, 95, 122, 124, 139
Stengers, Isabelle, 14, 53–54, 165
surgery, 20, 62, 64
surveillance, 16–17, 38, 39
SWIFT, 36

targets, 43, 124, 139; generation of, 5, 9, 16, 67, 90
technology, 15–16, 31, 36, 40, 43, 50, 101, 145
terrorism, 33, 55, 74, 89, 91, 124, 126, 138, 148
text, 93, 97, 101, 104
threshold, 41, 48, 54, 64, 68, 105, 139
time, 16–17

traces: data, 44–46, 62, 150

training datasets, 4, 9, 17–18, 20, 68, 96, 122, 125, 135–36, 150, 155; images used as, 49; in natural language processing, 90; patterns in, 54; and point clouds, 77

transparency, 5, 8, 18, 86, 110, 117, 123, 134, 166

truth: concept of, 5, 9, 14, 18, 93, 122, 135–37, 145–46, 148; and falsity, 103, 114. *See also* ground truth

Turing, Alan, 13, 56–57

Twitter, 89, 108–9

uncertainty, 12–13, 47–48, 55; machine learning and, 21, 22, 64, 88, 95, 111, 118, 136, 139, 142; political, 80; writing and, 98–99

the unknown, 90, 95, 97, 102, 105–7, 152–53, 155

US Department of Homeland Security (DHS), 2, 31, 43, 50

variability, 71–73, 126, 134

verticality, 17, 40, 46

violence: algorithmic, 4, 11, 46, 66, 118, 126, 146, 152, 161; meaning of, 91; potential for, 17, 111; science and, 134; sovereign, 53, 128, 164

the virtual, 35, 62, 177n15, 181–82n23

vision, 14, 39, 44, 53, 159, 162. *See also* sight

visualization, 14, 30, 38

war. *See* violence

weights: adjustment of, 24, 48, 68, 73, 75, 80, 92, 96, 103, 128, 145; and assigning value, 129; and indeterminacy, 22, 75, 147; machine learning and, 8, 13–14, 70–71, 75, 79, 106, 111, 128; and mediation, 162; parameters and, 74, 136; politics of, 163–64; and rejected alternatives, 81, 99

Wiener, Norbert, 111–15

Wilson, Charles Thomson Rees, 29–31, 44

Wittgenstein, Ludwig, 9, 13

world-making, 12, 20, 30, 33, 40, 43, 48, 88, 97, 161

writing, 86–88, 95–98, 101–3, 158. *See also* code

xenophobia, 119

YouTube, 109–10